CharcoalRemedies.com

The Complete Handbook of Medicinal Charcoal and Its Applications

by

john dinsley

Gatekeeper Books

Seventh printing June 2013

Remnant Publications
649 East Chicago Road
Coldwater MI 49036

Text Design by Greg Solie • AltamontGraphics.com

ISBN 978-0-9738464-0-9

Contents

Introduction

CharcoalRemedies.com is a book about charcoal. Charcoal is history, it is poetry and art, and it is science. Much more than this, charcoal is cleansing and healing for man and nature. Charcoal is a universe within a universe, and it is waiting for an invitation to yours. My acquaintance with this most unlikely of medicinal remedies was by surprise. Yours will be by design.

In 1897, health reformer E. White wrote, "I expect you will laugh at this; but if I could give this remedy some outlandish name that no one knew but myself, it would have greater influence." After all these years of knowing and using "this remedy", I find that this description of charcoal, better than anything else I could say, sums up the enigma of this lusterless all-time gem of medicine.

When I checked out from my "higher education" thirty odd years ago, shouldered my backpack and stuck out my thumb, I had my sights on the other side of my world. I was not alone. In the midst of the hippie era I met many young people my age crisscrossing the continents and oceans. I didn't reach my destination, but I did find what I was unconsciously looking for.

The 1970s were a time of exploration, of both outer and inner space. The notion that our persons were integrated dimensions and not just separate loose bolts was for many of us a welcome step back from the technocracy that seemed to reduce everyone to mere cogs in a giant wheel. Amongst all the movements during that era, there was a return to "mother earth", the more hands-on the better, including organic gardening, vegetarianism, herbs, natural fiber clothing, and holistic healing. So, when I wandered into a small medical clinic in Guatemala and discovered all that and more in a nutshell, well, I felt my spirit had found a safe harbor.

What I never expected to discover was how simple achieving good health can be, and how simple the remedies can be to repair broken health. And, if someone had told me that the cold remains of one of my many campfires held the antidote for hundreds, yes thousands of the world's ills, well, considering my frame of mind, I would probably have said, "far out!"

This book is first about that most unlikely of remedies, charcoal. The great Roman scientist Pliny wrote almost two thousand years ago: "It is only when ignited and quenched that charcoal itself acquires its characteristic powers, and only when it seems to have perished that it becomes endowed with greater virtue." But how would I have known that the very science and technology I was turning my back on, would, by the end of the 20th century, be absolutely and critically dependant on black dust? From Emergency Rooms to living rooms, from outer space to the bottoms of the oceans, from vegetable oils to cosmetics, from silk to art, from dialysis machines to kitchen water filters, from halitosis to hemorrhoids, from camels to roses, and from the living to the dead, charcoal touches each of us in hundreds of ways, every single day of our lives. How could I have been so unaware for so long?

Although my discovery of charcoal seemed accidental, the more I have studied, the more I have experimented with this simple natural medicine, the more I am convinced charcoal is no accident.

It is no exaggeration, this little book will open to you one of the biggest mysteries ever resurrected from the burial sands of ancient Egypt. This book will equip you with a working knowledge of the most powerful antidote for poisoning on planet earth. You will understand why it is the single most effective detoxifier of thousands of harmful chemicals anywhere in the world. You will be armed with a science that technology can dream about but never equal.

Are you a parent, a caregiver, a physician, a lover of animals, of flowers and the world around you? Do you work with your hands? Do you love exploring mysteries? Do you want to be healthy? Then this book has something for you. I know one who would agree. As a parent, a caregiver, a physician, and a lover of nature, Dr. Agatha Thrash also enjoys working with her hands, revels in the mysteries of science, and has made health a lifetime study. She writes:

> "Charcoal has amazing healing properties. In fact, if I were stranded on a desert island and could take only one thing along to protect me from disease, infection, and injury, I would choose charcoal." (2003)

Good mysteries always expose more mysteries, and one of the more troubling mysteries about charcoal is why so many people, perhaps yourself included, are still so uninformed about charcoal's amazing healing properties? When one considers that the fourth leading cause of death in America is due to the negative effects of **properly** prescribed drugs by doctors in hospitals,[1] thinking people should be asking, "Are there not safer alternatives?" There are, but…

Through all the passages and changes of my life, my appreciation for and fascination with the amazing qualities of charcoal has only increased. Quite the opposite from my appreciation, has been the general overall silence on the topic of charcoal therapy in the healthcare community. Apart from the odd reference, neither conventional drug medicine nor the different so-called holistic practices give charcoal the recognition it is due. While there are numerous essays and research papers on the wide range of uses of charcoal, there are only a handful of textbooks dedicated to charcoal as a remedy used for various health conditions. The most prominent authors are David O. Cooney, PhD, and Doctors Agatha and Calvin Thrash, MD.

Introduction

The late David O. Cooney was professor of Chemical Engineering at the University of Wyoming, Laramie, and was active in research on activated charcoal for medical applications as well as for uses in water purification. *The Annals of Pharmacotherapy* has cited his most comprehensive work, *Activated Charcoal in Medical Applications*, as "the standard printed work for information on the medicinal uses of activated charcoal ... an extraordinary book". It offers a wealth of clinical information.

Over the years the Doctors Thrash have published a number of textbooks (*Rx: Charcoal, Home Remedies, Natural Remedies*) on different health subjects. Salted through them are several stories about charcoal that they personally experienced or were told to them. They have graciously allowed me to include some of these anecdotes in this book. There are, however, no books available that tap into the thousands of personal experiences with charcoal that are being shared anecdotally. I felt, and others have agreed, that it would be a worthwhile venture to collect a sampling of these stories. The chapters that follow are full of history, science, recipes, resources, and numerous personal experiences (old and new).

Through the book the storytellers are introduced, for convenience, simply by their first names and by their profession. While charcoal is the natural remedy that is spotlighted, the reader will be reminded that there are no magic bullets, not even black ones. Good health, vibrant health, sustaining health is first a gift. We are privileged to have a vital part in maintaining that gift. Prevention, that is health maintenance, is the work of a lifetime. In the last chapters I try to capture some of the vital principles that underpin total health, that is the interplay between the physical, mental, emotional, and spiritual components of man.

When I embarked on this book I started by contacting old acquaintances for their experiences with charcoal. They inevitably led me to other people. The stories come from friends, family, strangers, doctors, truck drivers, nurses, chemists, musicians, housewives, ministers, tourist guides, vets, farmers, missionaries, businessmen, writers, and "your neighbor next door". I have grouped the stories around certain topics as they relate to humans, pets, gardening, industry, and the environment. I have included Chapter 4 for those of you who are intrigued with what is known about the mechanics and chemistry of charcoal. Chapter 17, very quickly looks at an assortment of other recognized natural remedies. *CharcoalRemedies.com* begins with a short biographical sketch, and concludes with a section on the holistic view of health.

Over the years and miles I have sat as a student and listened, and I have stood and lectured. I have been a patient as others have practiced their healing art on me, and I have bent over others applying my knowledge of simple home remedies. Over the months of compiling this manuscript I have found myself writing as the student, the teacher, the sufferer and the caregiver, and in no certain order. Those who I am indebted to for editing, recommended I lean to one audience. But I wonder if ultimately writers write for those who hold some common spirit. Whether it be the spirit of independence, the spirit of healing or helping, the spirit of empathy or sympathy, or the spirit of truth, the writer plumbs his audience for some note of resonance.

My intention, in assembling this information on charcoal, was to introduce its virtues and powers in practical ways. However, my initial world journey opened before me far more than the black, crusted edge of health. I dare not compare myself with the likes of Isaac Newton, or Albert Einstein, or Max Plank, or other great scientists, but like them my search for truth inevitably brought me to a crossroads. My world adventure turned into a spiritual journey that continues to this day.

As I camped alone by the sea or in the desert, as I gazed up into the starry heavens, I felt another voice plumbing the depths of my soul. I saw hidden within a new dimension. Plato once wondered, "Perhaps there is a pattern set up in the heavens for one who desires to see it, and having seen it, to find one in himself." I began to grasp the words of that champion of space, Werner von Braun: "One cannot be exposed to the law and order of the universe without concluding that there must be design and purpose behind it all. Through a closer look at creation we ought to gain a better knowledge of the Creator."[2]

CharcoalRemedies.com, you will find, is fact, but with a twist of lemon. It is story telling with hopeful endings. It is biting truth mixed with warm oil, a big smile ringed with black lips, big questions around a bigger Answer.

When all the pretense is stripped away, and all the traffic gone, do you see yourself much like everyone else, vulnerable, alone, small? If you have ever been through the fire of pain, or the sickness of sickness, and realized the brotherhood of humanity, then you will find yourself somewhere in these very brief windows into people's lives. If you have tasted the anxiety, desperation, and fear of trying to help some sick or dying loved one without enough knowledge or resources, you will be cheered with stories of heroes dressed in everyday clothes. If you sometimes feel overwhelmed by oppressive forces greater than your lonely self, you will be inspired with hope and faith and courage.

The study of charcoal, this simple benefactor of inner space, has further opened to me the great heart of the Creator, His personal care, not only for the universe but also for the sick and the forgotten. It is my hope, when you next find yourself by a fading campfire, as you rest your mind from contemplating the heavens above, as you look down at the glowing coals at your feet, that your heart and soul will be warmed with the knowledge that you have not been overlooked.

Introduction

CharcoalRemedies.com is about a nobody, a black sheep, who finds itself the benefactor of the poor and the rich. It is as comfortable around a campfire as seated in corporate boardrooms, ushered in by the most eminent of doctors and scientists. Like our little protagonist, when you feel as if you have been baked with unquenchable fire, broken and crunched, pressed out beyond measure, you too may find yourself ministering to the sick of body, of mind and of heart.

Disclaimer

I have been urged upon to include a disclaimer. I could only hope that those who read these pages will use the information wisely. I could only wish the application of charcoal, or the other remedies mentioned, would, in every instance, always bring about the effect desired. But I have learned there are also priceless lessons in pain and suffering and waiting, and God is too wise to err. I would therefore caution those of you who are prone to leap before you look, to take time before a crisis arises, to learn as much as possible about the bodies and minds you have been made stewards of. Then, knowing the science, apply the remedies with the seed of faith. CR

---

[1] Barbara Starfield, MD, *Journal American Medical Association* Vol. 284 July 26, 2000

[2] Werner von Braun (1912-1977) Rocket scientist and champion of space exploration.

A Word of Thanks

My sincere thanks to the many, many people who contributed the stories that are the inspiration behind this book. These stories turn otherwise flavorless knowledge and theory into real life drama. Due to space constraints, some of the individual experiences I collected were not included in the book, but will be posted on the website *CharcoalRemedies.com*.

Also, my thanks go to my sister, Deanna Robertson, whose editing skills helped to bridge the gap of understanding between what seemed so obvious to me, the author, and so obscure to the reader. I am reminded that in building anything with lasting impressions, there are the framers, and there are the finishers.

I am thankful too for my teachers, many of them being the sick and suffering who gave me permission to apply the simple home remedies that you will learn about in this book.

As with many authors, I must humbly thank my wife. Through this past winter, Kimberly literally kept the home fires burning.

An Invitation

If you find a blessing in these pages, or if you have proved charcoal to be a successful remedy for yourself or others, for your pets or livestock, for your garden or field, I invite you to submit your story to our website so others may benefit from your experience.

www.charcoalremedies.com

Chapter 1 A Little Bit of Black Magic

I still remember that warm lazy afternoon when the local bus stopped at the gate, and anxious parents rushed into the clinic with a barely conscious young child in their arms. The doctor, suspecting food poisoning, quickly mixed up a black substance and smeared the thick paste all over the girl's bare abdomen. Within fifteen minutes the pain and convulsions subsided, and the child was resting quietly. "Black magic! What next!" I thought.

It was twenty-nine years ago that I wandered into that small medical mission in the western highlands of Guatemala. After seeing a crumpled airplane in their small hanger, I was curious about the place, but mostly I was hoping for a free home-cooked meal. Doctor Graves didn't waste any time. He put me right to work—I was obviously not their first or their last hungry traveler. It wasn't until after a full day of labor that I would sit down to that home-cooked meal by a kerosene lamp. By then it was dawning on me how very different their lifestyle was from anything I was accustomed to. The food was plentiful and nourishing, and the spiritual atmosphere was just what I had been praying for. So I asked if I could stay and work for my board.

This small one-doctor medical/dental clinic was as busy as the beehives kept to supplement the mission's table and its income. From morning light till after dusk, patients could be seen waiting outside the clinic hoping to see the doctor. Each day seemed to expose some new dynamic of Clinica Valparaiso. Early morning devotions, singing and prayer, a vegetarian diet, an elementary school, live-in patients, evening worships, and a network of Bible workers crisscrossing the mountains and valleys, were all integral parts of the clinic and the mission. Like their beehives, the mission hummed.

I was intrigued with the whole place, but it was the emphasis on disease prevention and health education, as well as home remedies, that was so novel to me. Dr. Graves, a U.S. board certified MD, was ever ready to administer or suggest some simple natural medicine to his patients, even though he had a dispensary full of powerful modern drugs to choose from. And he practiced what he preached. For some time he had had a cancerous condition, which, he told me, he had been able to contain with simple remedies.

So, what was it that Dr. Graves used to so quickly and easily rid that young child of the poisons in her body? What was it that was so simple, so effective, so universal, so inexpensive? Charcoal. Yes, simple charred wood. This was my first (but not last) encounter with the wonder-working power of this natural remedy that dates back to the mysteries of ancient Egypt. From the jungles of Guatemala, to the steps of the great Himalayas, from the coral atolls of the Pacific to the concrete jungles and towers of the 21st century, I have again and again marveled as I have witnessed the efficacy of this humble antidote of poison and disease.

As I became involved in overseas development work, and was exposed to new environments and a host of mutating pathogens, I have many times been rescued from the ravages of sickness and disease all around me. I have for almost thirty years, carried a bottle of activated charcoal wherever my travels have taken me. It has been a first line of prevention and treatment. Over the passing years, not only has it served as doctor and nurse for me, but also for many of the sick and suffering I have met along the way. However, before I go on to explore with you the wonders of this unlikely simple remedy, let me sketch some of my travels before and after this juncture in my life that helped shape my interest in a holistic approach to health.

My Quest

Several years earlier, I had taken a general science program at a university, but my spirit soon tired of it. After a couple of years, I checked out of my "higher education" and headed to the high arctic to make my fortune working on the oil rigs. After a few more years and some near fatal accidents, the prospect of being dead with a fist full of money seemed even less appealing than university. I decided I was ready to resume my education.

My father had cultivated a love of nature in all his children. When any of us had an opportunity, we loved to travel to those places we'd read about in books. My older siblings had already ventured away from their roots, and at last I felt it was my turn. Our mother had given us a love for knowledge. We were equipped to think and to research, but I wanted to study in a less structured environment. So I loaded up my backpack and hit the road. After a year of crisscrossing America, I decided to head down through Central and South America, and then somehow hop over to New Zealand.

When I had first started off on my journey, my parents surprised me with the gift of a small Bible. It seemed odd since I never pictured us as a religious family. My father was agnostic in his belief. My mother had ushered me off to Sunday school up until I was sixteen, but our attendance had never been a regular thing. My brush with higher education had buried any religious inclinations I may have had. How unexpected then was this gift.

Over the years I had again and again been accosted by different ones claiming to know what the Bible said about this and about that. I decided to read it for myself. I wanted to be able to answer intelligently when confronted by others. So I tucked the book into my bulging pack and stuck out my thumb.

Each day before venturing out to the highway, I devoted some time to reading a couple of chapters from my newest "novel". I am from a family that devours books. But this one ground me down to a snail's pace. If biochemistry and calculus textbooks wearied my brain, the Bible made my head ache. But I persisted, and as the miles slipped away, along with more and more things from my pack, I began to feel a lot lighter, and not just physically. My fascination with psychedelic drugs began to dissolve, and my taste for alcoholic beverages ran dry. As well, other practices I never consciously tried to modify seemed to quietly fade out of my life. I also saw myself rescued from events and in ways that defied the odds.

By the time I got to Mexico, I had begun to very privately explore the possibility that if there was a God maybe He could hear me. Sitting beside a campfire on some secluded beach, on a hill, or out in the desert, in my own way I began to converse out loud, not really knowing if I was alone or not.

After a month of hiking, sometimes a hundred miles from the nearest pavement, I emerged unscathed from my desert experience, only to be robbed of my passport. The next day I got sick, very sick. But somehow, with each new obstacle, things always worked out better than I could have possibly imagined. And so it was that one day I found myself walking up to that small clinic in Guatemala.

Belize

After several months under the care of the Graves family, I had gained about ten pounds and was feeling great. Great that is except for another case of itchy feet. I had barely ventured off their property during my time there, and, once again, I was anxious to continue on with my travels.

Due to visa permit restrictions, the doctor's son needed to exit the country. This time it would have to be by some other way than by plane. Some months previous, Dick had brought their small Cessna airplane down from California on its maiden flight. As he wound his way down into the tight valley to make his landing, a Highways Department truck pulled onto the airstrip next to the highway, and Dick was forced to park the plane in the trees on an adjacent hillside.

Dick was quite the adventurer too, so we decided to travel together cross-country to another clinic in Belize. We hiked, rode on top of a bus, bounced down miles of washboard roads to the road's end, and then paid passage on a motorized thirty-foot dugout canoe. We got stranded half way to the next road connection and ended up buying a very tippy eight-foot dugout, with a board and broken paddle thrown in. Between laughing, capsizing the canoe, sweating, and staying with local families along the way, we finished the rest of our hundred-mile journey down the river.

With each stop Dick became instantly popular. Popular, not just because of his red beard and good Spanish, but also because he happened to have brought along some dental pliers. Once people found out about his dental skills, they lined up to be his next patient. Each day we paddled a little farther.

Surrounded by all that water, we had a hard time finding any that was drinkable. We had to treat what we had with iodine. If I had only known a little more about charcoal and what it can do to foul-tasting water! By the week's end we were no longer laughing, and were very glad to sell the canoe and buy some bus tickets. La Loma Luz clinic in Belize was our next welcome retreat.

I would end up staying in Belize for the next year and a half. Unfortunately, Dick became very sick with hepatitis and had to fly back to Guatemala. If I had only known then what I know now about charcoal and its use in treating hepatitis!

In Belize I was once again exposed to all the changes I had been introduced to in Guatemala, and more. La Loma Luz ("The Hilltop Light") was a newer undertaking, so in some respects the conditions were more primitive. When Dr. Mundall learned of my building experience, he asked if I would help with the construction of their new clinic hospital. I was moved by the willingness of this family to leave their very prosperous practice in Arizona and come to such primitive circumstances. So, I agreed to stay and help. Together with the doctor's twelve-year-old son and two local boys, we began to pour the footings for a 110 ft. x 90 ft. building. Having only shovels and a wheel barrel to begin with, the project seemed more than a little overwhelming.

Over the next year and a half, except for the couple months I was away working as a cowboy on a large cattle ranch, I watched the clinic walls slowly go up as different volunteer workers came and went. I had made a silent promise that I would stay until the roof was on. It never dawned on me that one morning the doctor's wife would announce she was strongly impressed that the new building needed to have a second floor. I stayed on, but the day that last sheet of roofing was nailed down, I packed a few things into a small Guatemala bag and said goodbye.

During my time there in Belize, I had been exposed to malaria, infected with worms, contracted a severe case of amoebic dysentery, and lost all the weight I had gained and more. But I did not feel at all discouraged. I could look back and see what a tremendous education I had received. My life had been radically changed. I could never go back to seeing things the same way. I had seen people being helped with all types of problems.

I understood the need to take more responsibility for my own health and not leave it to chance. I knew the importance of prevention first hand. I saw the place for emergency care and the power of simple remedies and careful nursing. And my spiritual nature had come into sharper focus. I felt an uncomfortable responsibility to those less fortunate, that I was somehow indebted to the poor and should stay on and help. Despite all this, with twenty dollars in my pocket, I finally decided to continue on with my round-the-world adventure.

A New Direction

Now here I was once again, after a year and a half hiatus from my travels, ready to pick up where I had ended, and head south on my own. But somewhere along the road I was struck with a strong sense of mission. I continued on as far as El Salvador. There, I made a U-turn and headed back to Clinica Valparaiso.

As I made my way north to Guatemala, I found the spiritual dimension of my life opening wider. I now felt comfortable in making public my belief in God, and I was anxious to know more. My mind and heart echoed the words of Albert Einstein after he became converted to the notion of a Creator God: "I am not interested in this or that phenomenon, in the spectrum of this or that element. I want to know His thoughts; the rest are details."

It seemed fitting to mark this transition in my life with a "burial" of the old me and the metamorphosis to the new. I was not the only one who had had a reawakening in the past two years. On my arrival back at Clinica Valparaiso, after some discussion with the doctor, the two of us drove to another town, up in the mountains near Antigua, and we were baptized in a large communal water trough.

The doctor, seeing my new interest in health and in serving others, offered to train me in every aspect of his work, and if I would stay and help him, to give me half of the earnings from the clinic. I was very humbled by the offer and, except for some unfinished business I needed to attend to back in the States, I was inclined to stay. As it was, I shouldered my bag and continued on north. Little did I know that my sabbatical from formal schooling was drawing to a close. A whole new chapter of health, healing and simple remedies was preparing to open before me. CR

Chapter 2 Charcoal Therapy

Someone once told me, "There is no such thing as coincidence. Behind all the threads of our lives there is a master design." The amazing sequence of events that eventually brought me to a small health training facility in Alabama confirmed to me the truth of these words.

Holistic Medicine

Calvin Thrash, MD, a board-certified specialist in internal medicine, and his wife, Agatha Thrash, MD, a board-certified specialist in pathology, started Uchee Pines Institute thirty-four years ago. After years of practicing conventional drug therapy, they experienced a fundamental change in their philosophy of health and consequently developed an educational and treatment program based on a holistic approach to treating disease. While the courses were not accredited, they were just what I was looking for. They offered a comprehensive, and concentrated training that suited my plans. I wanted to return to Central America and become involved in some form of development work.

At Uchee Pines there was a strong emphasis on prevention, and on public health education. The many patients who came were treated with simple home remedies, and vegetarianism was a major component of the program offered to them. The students helped with growing and cooking most of the fruits and vegetables placed on the tables at mealtimes.

Students, patients and teachers were involved together in a wide range of outdoor activities. Sunshine, fresh air, water, exercise, and rest were considered nature's most treasured remedies. As part of their public health practicum, the students gave health lectures on a host of different topics to the lifestyle guests. Cooking schools, and stop smoking programs, were also made available to the community through seminars.

There were daily classes in physiology, anatomy, nutrition, standard nursing practices, and more. We all attended with the doctor as she performed autopsies, and the women helped with home births. In fact, the education was quite comparable to a physician's assistant training program.

The actual hands-on treatments revolved mostly around different hydrotherapies, but also included physiotherapy, massage therapy, and the application of poultices.

As simple and practical as the physical health components of the program were, so was the attention to the spiritual dimension of healing. Prayer was encouraged along with sound health principles. The belief being that God's desire was to bestow health and healing as patients cooperated with the Creator's laws of health.

Over the course of the two years that I attended Uchee Pines, I learned how to take a medical history, do physical exams, draw blood for blood work, give a broad range of rational treatments, prepare diet-specific meals, give health talks, conduct public health programs, and run a bakery. It was also at Uchee Pines that my catalogue of medicinal uses for charcoal began to grow.

Charcoal Therapy

Charcoal was given internally to help detoxify individuals for a variety of toxic conditions, and it was also used externally in poultices. While different ingredients were used to make poultices, there was a clear preference for charcoal. What I had only witnessed casually while working in Central America, I now was able to see applied on a daily basis to a broad range of diseases and conditions. One of the more common uses of charcoal is to control flatulence, and that was really the limit of my own occasional experience with it. However, while at Uchee Pines I was able to observe some amazing applications of charcoal both internally and externally for other conditions.

I found some of the charcoal treatments truly amazing. The memory of one treatment, given to a gentleman with liver cancer, is still vivid. His abdomen was quite distended from his condition, making it difficult for him to breathe. I had not been assigned to work with him, but stepped in for one of the other students. I followed the doctor's orders, and made a charcoal poultice for the patient to wear over the kidney area on his back during the night. The next morning when I removed the poultice, I was completely incredulous at what I discovered. There was a distinct yellowish stain in the white areas of the poultice, and even more amazing, the poultice reeked with that unmistakable odor of urine! It was clear to me that the charcoal poultice, in some mysterious way, had managed to help eliminate body wastes. When I later learned that kidney and liver dialysis equipment use a bed of charcoal through which the blood is filtered, I better understood why the doctor had prescribed the poultice.

Previously I had heard the doctors and others relate what I was sure were exaggerations concerning the results they had seen with charcoal. This was the first time I had been directly involved, and I knew the certainty of what I had observed. That was the limit of my involvement with that patient, and I don't recall the outcome of his stay. However, it wasn't long before I would have the opportunity to see the amazing healing qualities of charcoal again and again first hand.

On completing our course of study, my new wife and I headed for Puerto Rico. There we joined a group of local families hoping to develop a training facility patterned somewhat after Uchee Pines. We lived and worked with this group for over two and a half years.

Since then my travels have taken me back and forth across America, to Asia, to islands in the Pacific Ocean, and back again. However, through all the passages and changes of my life, my appreciation for, and fascination with the amazing qualities of charcoal have only increased.

Whether you are one of the fifty million North Americans who cannot afford health insurance, or one of those who pays large premiums for health coverage, you cannot afford to pass by this modern day wonder that is as ancient as the mysteries of Egypt.

Charcoal, as common and as glamorous as the bathroom toilet and yet less understood, less recognized, and less appreciated. Charcoal: unassuming, universal, international, cosmopolitan, inconspicuous, impartial, industrious, indispensable, benign, benefactor, mysterious, and black. The black sheep to some. The black diamond to others.

How could something so widely exploited in this technological age be so under-utilized in the fields of disease prevention and cure? Curious? Join me as we follow this globe-circling ambassador as it speaks to the power of nature's simplest remedies. *CR*

Chapter 3 How Do *You* Spell Relief

There is a time for everything, a time to study, and a time to take what one has learned and put it to practical use. My quest for knowledge had certainly redirected my adventures, but it had not ended them. My two years at Uchee Pines had given me a working knowledge of the human body, a grasp of the principles that underpin good health, and an experience, albeit small, in applying an assortment of different simple natural remedies, one of them being charcoal. But, I was anxious to be off and on our own. In hindsight, our leaving Uchee Pines was premature. Nevertheless we purchased plane tickets and left.

It is one thing to assist the sick in recovering their health when you are under the watchful care of seasoned instructors. It is quite another thing to venture out on one's own. However, it is just that sense of insecurity, of being thrown outside of our little box, that gives one experiences that no amount of instructing will ever equal. That is not to say that we no longer need instruction. Rather, it is our own practical experiences that give us a passion for what we have only learned from books. As well, it is our own practical experiences that give us credibility.

This chapter recalls some of my experiences that have included charcoal. I don't want to give the impression that charcoal is the only form of treatment that I recommend or use when I am called upon to help the sick. It is not always the first choice, but neither is it the last. It is one remedy among several. However, each case of suffering that has been relieved using charcoal, including my own, has galvanized my faith in its ability to detoxify the body of any number of poisons. My hope is, as you brush up against these stories of mine and of others, they will inspire you to try charcoal for yourself.

Puerto Rico

Having completed our studies at Uchee Pines, my wife and I set off for Puerto Rico. We took up our work in the hills of Galateo, three kilometers up and down a four-wheel-drive road. No electricity. No running water. Primitive conditions, but what spectacular scenery! Our long busy days were spent building a health center, clearing land and carving out terraced gardens on the steep hillsides.

For convenience, we and the local families we worked with, all ate together. The women prepared simple meals from local products. The food, although nourishing, was new to us. One of the favorite native dishes was beans—three times a day. I like beans, but, as they do for many others, they can inflate me a size larger. As the months passed my enthusiasm at mealtime became more and more strained. Fortunately, I always knew I had one ally. By swallowing a couple of charcoal tablets or a tablespoon of powder in water, within the half-hour I would once again be able to tighten my suspenders.

It was not long before we began to share with the group of families, the health principles we had learned. We demonstrated lifestyle practices (Chapter 19) and simple treatments (Chapter 17) that produced positive changes in their health. Soon we began to have a slow but steady stream of visitors hiking into our little community. Some came out of curiosity, some to listen to our talks, and some to ask questions about matters of health.

Foot Ulcer

I remember well one early request for our help. I was approached by a man who was very concerned for his elderly mother. He described an open lesion the size of a silver dollar on her foot. She had been attended by different physicians, but the condition had not responded to conventional medicine. I suggested that he have his mother soak her foot twice a day in a charcoal foot bath with enough water to cover the ulcer. After inquiring into her eating habits, I also suggested some dietary changes that I felt would improve her circulation.

A month later when the man brought his mother to meet us, the ulcer had shrunk to the size of a quarter. However, we noticed that her legs were dark blue from the knees down, a condition that her son had failed to mention during his first visit. The son asked if his mother could stay with us. Because of our primitive conditions, we were not well prepared to take her in, but we wanted to help. By the end of a month, with daily charcoal baths for the feet, the sore had completely healed over. Equally important, by maintaining a strict low fat diet of whole grains, fruits and vegetables, both her legs regained good circulation and their normal color.

At that time, because charcoal was still used locally for cooking, it was easy to purchase a large lumpy burlap sack full. With a little work, the large chunks could be pulverized into a crude powder. It wasn't the finely powdered activated charcoal we were accustomed to, but it worked. That is what was used for the woman in her home, and not the commercial powder we normally used.

There is a joy in seeing those who have been seriously ill, restored to health after years of sickness. Too often chronic suffering stems from years of small health indiscretions. People then resort to drugs, hoping to reverse their previous disregard of nature's laws. Treating the immediate symptoms of the disease with the use of nature's benign remedies is far safer and kinder than further weighing the body down with drugs and their adverse affects. After achieving a measure of relief from pain or discomfort by the application of some simple remedy, the attention can then turn to understanding the cause of the disease. The patient should be encouraged to cooperate by changing any unhealthy habits that might aggravate their condition. In this way improvement will come naturally. Simple remedies will not tax the system as the general health slowly recovers. Charcoal is one such rational treatment.

Hepatitis

Another of our earlier and rewarding experiences with charcoal, had to do with hepatitis. Pastor Caban suffered with chronic hepatitis and had been released from his ministerial duties. He had been in and out, and back and forth to the hospital several times. However, his condition was not improved. He had heard that we had been instrumental in helping different people suffering with various ills. Their health had responded favorably to some small changes in their lifestyle along with some simple treatments. But the minister was not convinced. He had, like most of us, heard of some bizarre treatments and fly-by-night clinics. But when some relatives reminded him that he had not improved with conventional therapies, he decided he had little to lose by coming and staying with us.

It is strange to see a dark skinned person with a yellow tinge. Pastor Caban's jaundiced condition was most obvious in his glazed yellow eyes. This yellow pigmentation is a sure sign that the liver is not doing its job of filtering out certain waste products in the blood. It so happens that direct sunlight is able to break down these bile pigments that accumulate in the skin. So, besides having him follow a simple vegetarian diet, we scheduled Pastor Caban to take a sunbath twice a day. He found both these remedies to his liking.

There is an age-old maxim that bears repeating, "The life is in the blood". It would follow then, that perfect health would depend upon perfect circulation. This is the most fundamental physiological principle to be considered when striving for optimum health. With this in mind, we began applying moist hot packs (Chapter 17) over his liver. This would naturally promote circulation and stimulate his autoimmune system. These fomentation treatments, as they are called, can also be very relaxing to the patient. Pastor Caban did not seem to mind these at all. Having dispelled some of his apprehensions, he was more receptive to our next suggestion.

Knowing charcoal has a strong affinity for bile products, we now asked him to begin taking activated charcoal daily. He was to use it both internally and as a poultice over his liver at night. The charcoal would help to relieve some of the workload on the liver. Waste products within the bowels would attach to the charcoal and toxins in the blood would be drawn through the skin into the poultice.

Whether it was the sunbaths, the diet, the fomentations, or the charcoal, I cannot say, but I can vouch, as can Pastor Caban, that within three weeks not only did his color return to normal, so did his liver functions. The latter was confirmed by blood work done at the hospital. Incidentally, Pastor Caban also suffered with hypoglycemia. As expected, when he began a strict diet of simple whole fruits, grains, and vegetables, with no snacking between meals, his symptoms quickly disappeared.

It was not long before Pastor Caban was able to return to his three churches. And he returned to his work with a mission—to preach a message of health in conjunction with the simple agencies of nature.

While not all experiences will have so dramatic a turn around, I have found that, if the pain and discomfort that come with disease can be lessened, the simple remedy has been a success.

Water Filter

Charcoal is a simple remedy in its own right, but it is also a remedy of remedies. Charcoal is also doctor to some of the food we eat, the air we breathe, and the water we drink.

"Water, water everywhere and not a drop to drink", was a favorite line my mother often quoted. Thankfully that was not one of our worries out in the hills of Galateo. At the foot of our Mt. Galaad, we had a spring of the freshest, tastiest water anywhere. But that was not the case everywhere on the island.

On one occasion my wife and I visited a friend in a distant city notorious for its dreadful water. We had forgotten to bring our own water with us. Driven by heat and thirst, I was at last forced to ask our host for some of his. I was directed to a beaker in the fridge. The water was clearly gray! "Attitude" kicked in. My friend, however, assured me it was okay. Since he was an electrical engineer for the city, I had to believe him. I do not exaggerate. The water was as pure-tasting as our mountain spring water. In another beaker, from which the gray water was poured, was one of those large unappealing chunks of charcoal. Today, twenty-six years later, homes everywhere display an array of shiny chrome charcoal water filters.

Food Poisoning

It is nice to know charcoal will adsorb thousands of pollutants as well as different microbes. Unfortunately it is always easy to be careless with hygiene, especially in warmer climates. So it was for us one particular Sabbath afternoon. We had eaten a delicious lunch, and later went for a leisurely walk along the road under the spreading tropical trees. As we looked out over the relaxing pastoral scene, my stomach began to feel less and less restful. By nightfall I was feeling far beyond uncomfortable. I took some charcoal tablets, but I knew that this time it was not just gas. Midnight came and the pain was so intense that I could no longer lie still. Vomiting turned to retching, and I finally ended up on the floor in a fetal position. I don't know why it took us so long to think of it, but at length my wife mixed up some charcoal and made a poultice. After she laid it on my abdomen we prayed. The pain became so intense all I could do was hug the poultice as I rolled on the floor.

By 3 a.m. I had had enough of pain, simple remedies, and prayer. "Take me to emergency," I groaned. "I can't take this." However, before we left for the hospital, I decided to try a poultice once more. Another one was made, only this time it was very wet. We put that on my abdomen and said another prayer. I lay back on the bed and within five minutes I could feel the cramps easing. Within the half hour, I was sleeping. When I finally got up later that morning, my stomach was still sore from all the convulsing, but the poison, whatever it had been, was neutralized. I was a confirmed believer in charcoal. It really DID work!

Jaundice

It's one thing to believe, and to practice some remedy on oneself. It is something else again to care for a child. When Nathan, our firstborn, came along, he was somewhat jaundiced. The yellow-orangish appearance of his skin and eyes was due to the build up of bilirubin, a bile pigment that was not being properly metabolized. For various reasons, the liver sometimes does not kick into gear at birth, as it should have with Nathan. Out he went into the sun for a daily sunbath. Charcoal has also been credited with lowering bilirubin levels. But, since babies are only designed to swallow at birth and not chew, we mixed some activated charcoal powder in a bottle of water and let the particles settle out. We then poured this gray water off into a baby bottle and popped that into his mouth. After a couple of days, and several ounces of slurry water later, he was a healthy ruddy pink.

As he grew, Nathan would now and again show signs of being a little colicky. We could only smile as he would accept a charcoal tablet, and then thoroughly enjoy playing with it in his mouth. By the next morning he would be over whatever had caused him discomfort. Later, when his brother Enoch came along, charcoal tablets were his first experience with "candy." If only other young parents knew how powerful charcoal can be as a first aid.

Many miles and years have passed. Children grow up, their parents grow older, but charcoal never ages. Literally. If stored dry, away from heat in a sealed jar, charcoal will retain its adsorptive qualities indefinitely.

Nepal

Several years ago, thousands of miles away from the balmy climate of the Caribbean, I found myself looking over another perfect little baby boy. Only he was lifeless. The child had swallowed some kerosene. I had joined a group of teachers and students at a remote clinic in the foothills of the great Himalayas. Learning some Nepalese as I went along, I helped construct a water system for several villages and also for the clinic. The others helped by teaching English, nutrition, hygiene, simple treatments and organic gardening. Because of my experience, I was often called upon to attend to some of the sick who came when we were without a nurse. Some of the cases were sad indeed. How does one hand back to a young father his lifeless baby boy? With help so far away, if only the family had come sooner, or if only they had known to first try pulverizing some charcoal from their fireplace, the outcome may have been so different. As a last resort and somewhat poorly, charcoal will adsorb kerosene from the bowels when given in large enough quantities.

Abscess

I think the biggest thing foreigners remember about Nepal, apart from the sky-scraping mountains, is the little children. They are everywhere, looking out from behind a bush, watching you from a distant hillside, or poking a hole in the plastic covering the window so they can peek inside. One grandmother brought her little grandson for me to see. I had noticed him often. He was not mischievous like some of the others. He was shy and quiet, and he was obviously in pain. His grandmother pointed to the bandage on his leg. The local community health worker had attempted to pack the hole in his calf with gauze. I assumed the boy was supposed to go back to have it redressed, but he had not gone, and the sore had abscessed. It was quite infected and smelled terrible. As I removed the bandage, then the gauze, the tissue sloughed off, leaving a one and a half inch deep hole. After cleansing the wound, I packed it with activated charcoal and placed a charcoal poultice over that. He was told to come back the next day, and he did. With a new dressing each day, the wound quickly healed. The grandmother was most grateful.

Kiribati

The following year I joined a small organization working in the Republic of Kiribati, formerly the Gilbert Islands—a string of coral atolls straddling the equator in the middle of the Pacific Ocean. Using a small motorized sailboat, the workers were ferried from island to island to do medical and dental clinics. I stayed on the island of Abemama for half a year teaching organic gardening at a boarding academy.

On our way there, we first stopped in at the small island of Kuria. When we gathered that evening under the maniaba (a large coconut thatch shelter), the local minister joined us, limping as he came. His face grimaced in obvious pain. He struggled to focus his thoughts, but it was clear that he was in great discomfort. One foot throbbed with such pain, he said, that he had not slept for weeks. His big toe was swollen, reddish, and especially painful. From what he described, and since he had no history of injury, I wondered if he was suffering from gout, a condition in which uric acid crystals form in the extremities.

Knowing that the sick are seldom able to take their own recovery in hand, I decided not to wait till morning. I went back down the trail to where I had seen some burnt underbrush, gathered some charred branches, and brought them back. I scraped the charcoal powder into a basin large enough for his foot, and added warm water. For convenience, I also gave him several activated charcoal capsules to take internally.

Next morning, as I approached the pastor's compound, I found him scraping charcoal into a basin. The pain had been relieved and he had slept. You can imagine his gratitude. He was now prepared to hear how he could modify his diet so as to minimize the production of uric acid.[1]

After leaving the island of Kuria, we anchored in the sheltered coral bay of Maiana. One could only suppose the waters to be the purest anywhere in the Pacific. But various diseases have found their way across the distances, and contaminated the fragile water table, and even the bays. But I didn't know that when I later waded back to the sailboat with a badly stubbed toe. As we visited with the government nurse stationed on the island, I mentioned the use of charcoal for various problems, and asked if she had ever had occasion to use it. "Oh yes!" she replied. "A while back, when all my conventional drugs had failed to help in a severe case of cholera, as a last resort, I administered charcoal and the patient fully recovered."

Unfortunately not all cases of cholera end on such a good note. The year before, not long after I left Nepal, I was saddened to hear that an acquaintance who lived in a nearby village had quickly succumbed to cholera. He left his little family without a husband and father. If his relatives had only known about charcoal, maybe the outcome would have been different. When I left Nepal, I had counted myself so fortunate that, unlike the other workers, apart from a little diarrhea, I had not gotten sick.

Bacterial Infection

I was destined to have a very different experience over the course of my time in the South Pacific. Not long after we had arrived at the island of Abemama, my home for the next five months, my stubbed toe became very inflamed. I didn't think too much about it. How bad can a stubbed toe get? Well it almost claimed my life. The bacterial infection traveled up the lymphatics of my leg and erupted in boils above the knee. I watched as the infection proceeded to eat out small craters of flesh, muscle and sinew. One hole was a good two inches in diameter by an inch deep. As providence would have it, I remained behind alone when the others sailed on to another island. I ran a raging fever for days. One night, with my only comfort being the flickering candlelight, I made my peace with God, sure that I would not survive the night. But the fever peaked, and the next morning I thought the worst was over.

The days following, using a chair to support myself, I would limp from my little thatch cabin to our cookhouse. There I was able to make hot packs to put on my leg. After my treatment, I would make a charcoal poultice. Retreating to my cabin, I would cover the worst sores with the poultice, and then lie down again. The pain from my boils was the most intense pain I have ever experienced. The charcoal did not seem to help the pain, but it did speed up the growth of new tissue. The open sores soon healed over. Then I understood where those nasty scars on the native folk had come from. I was so glad to be on the mend.

Unfortunately, I was still very naive about protecting my feet. By wearing only thongs so as to expose my toe to sunshine and air, I set myself up to puncture the other foot with a stick. This time I immediately gave myself a charcoal foot bath. The puncture healed much more quickly. Having since gathered stories from others, if I were faced with a similar case of boils, I would now sprinkle dry charcoal powder directly into the oozing wounds or pack the open ulcer with charcoal jelly.

Cape Breton—Ulcers

When I returned to Canada from the Pacific, I decided to move to Cape Breton, Nova Scotia. I purchased this old farmhouse and began some major renovation work. One day I needed a tool for a job and headed off to the nearest tool rental store. As we were loading the truck, I could tell that the young attendant who was helping me was in a lot of discomfort. Seeing the strain in his face, I asked if something was bothering him. He said he was suffering with ulcers.

In his mid-twenties, John's duodenal ulcers had become so bad that they were affecting his marriage. After admitting he drank coffee, I explained that coffee dramatically increases the acid production in the stomach. He said that if he had to stop his coffee then he would live with the ulcers. "Well" I said, "then you can think about having eroding ulcers". That grabbed his attention, so I told him about charcoal, and how it works amazingly well to neutralize stomach acidity. Skepticism spread across his face with that typical Cape Breton look that says, "Go 'way!" Hoping it would inspire a degree of confidence in my unusual remedy, I directed him to the pharmacy, where they could order the charcoal, if there wasn't any in stock.

Then I realized that his going to the drugstore was not going to happen. Just for such skeptical folks, I carry a small bottle of charcoal capsules in the truck. I pulled it out, gave it to him with some simple instructions, and said goodbye.

I didn't see him until the following week. With a big grin, he immediately announced, "I am totally free". All his symptoms were gone. I could tell just by looking at his face. Keep in mind that John had had these severe pains for quite some time. I then had a chance to mention other items he needed to be careful with, such as spices, condiments, and smoking.

I saw John a month or two later and he said in a very confident manner that not only had he stopped coffee, but that he had also stopped smoking. Now, if he were to feel some acid indigestion coming on, what do you think he will reach for, some brand name antacid with calcium, which actually increases acid production? No, like many others, John knows 'relief' can also be spelled:

C-H-A-R-C-O-A-L

Gallstones (?)

My next-door neighbors, here in Cape Breton, are dairy farmers. They are hard working and work long hours. They seem to never stop. One day Michael was wearing one of those faces that lets you know that the person is having a rough day. He told me his long-standing problem was back for a visit. Michael, on a semi-regular basis, experienced severe indigestion and pain, which left him very ill for three or four days. He had been to the hospital several times for x-rays, but they showed nothing abnormal. He would feel nauseous, and then vomit, have diarrhea, and pain.

This had persisted for three years. Each time he went to his physician nothing conclusive was found, and he would leave with a popular drug to short circuit his nausea and vomiting, and some painkillers. Then he would come down with it again, and be laid up for three to four days. "I knew when I was going to be sick, because I would wake up with a taste of eggs (sulfur) in my breath." He would get up, go out to feed his cows, and then by the time he returned for breakfast, a couple of hours later, he would be feeling very sick.

I suggested that he try some charcoal. I gave him a number of tablets, and told him to take several when he was not feeling well. He was to continue taking them throughout the day as long as the symptoms persisted. I told him I was not sure if it would help with the pain, but it should help the stomach upset and diarrhea.

Michael did take the charcoal, and that was the end of his routine of three to four days of pain, vomiting and diarrhea. It was not long before the tablets ran out, and he asked the pharmacy to stock some. Whenever he woke up with the familiar "egg taste" in his mouth, he would immediately take two or three tablets.

Michael remained symptom-free, except for the occasional egg taste, for the next two and a half years. His doctor insisted that the charcoal was not doing anything. However, he said that if Michael thought they were doing him good, then there was nothing to keep him from taking them. The charcoal would, his doctor said, do him no harm.

A new doctor moved in just down the road from us and explained to Michael what he thought had been happening. Michael was, he said, probably passing gallstones, followed by a rush of bile into his stomach. This would explain the egg taste, nausea and vomiting that Michael had been experiencing. Because Michael always showed up for x-rays after the stones were passed, they were never detected.

After two and a half years of avoiding another trip to the hospital, Michael woke up one day with severe pain. The charcoal did not help. He was hospitalized, and he did have gallstones. When they removed the gallbladder and bile duct, it was evident from the scarring that he had been passing stones for several years.

While charcoal did not cure the cause of Michael's suffering, he would be the first one to say that it worked dramatically better than anti-nauseants and painkillers.

Promoting Charcoal

Charcoal can be a hard sell. If, when trying to introduce charcoal, I meet hesitation or apprehension, I can only reflect back on my own initial skepticism. My brother Ron has not always "appreciated" the value of my health innovations, but today he has no doubt as to charcoal's benefit when it comes to stomach upset. As Canadian west-coast commercial fishermen, he and his wife Gerrie worked long and hard seven days a week during fishing season. "There was no time to prepare a proper meal, so then we ended up having a big meal just before going to bed. That habit tended to end in pain, with an upset stomach and gas. But we would take several capsules of charcoal, and the symptoms promptly disappeared." Back on land again, they are careful to better plan their mealtimes.

Revitalized

A gift of health can be like the cup of water that primes the pump. Many who are sick have no idea where to begin when they have lost their health and their will to regain it. But once inspired with a measure of renewed energy, some, who recover from their sickness, will begin to take a more holistic look at their health. Hopefully these pages will do that for you.

Your initial reaction may be to discount these stories as coincidental. Some may brand them as quackery, and never give them another thought. How could charcoal be beneficial in such a wide variety of settings? Not unlike Pastor Caban, or John, or Michael or Ron, most of us would say, "Prove it to me and then I'll believe you." On that invitation, I challenge the reader to look objectively at the facts. The following section explores some of the history and science behind charcoal—the mechanics and the mysteries.

But first, let's retrace some history and see how science and industry have teamed up to put this little Trojan horse to work. CR

---

[1] Uric acid is metabolized from foods high in protein, particularly animal products.

Chapter 4 Black Holes

Outer space is a constant mystery. Of all the wonders of interstellar space, invisible black holes are a gigantic mystery. Though no one has ever seen a black hole, physicists have for many years insisted they exist. Just as we gain a better understanding of the invisible wind as it bends the trees or twists dirt devils in the desert sands, in the same way, events happening around black holes allow scientists to understand a lot about them without actually seeing one. As an analogy, let's compare one giant mystery, a black hole, to a very tiny one, a charcoal particle.

We have been told that black holes in space are actually intense electromagnetic fields that capture anything within range. Once caught, matter/energy is, for all practical purposes, entrapped. If nothing else, it sounds fascinating. Now who would guess that the humble little grain of charcoal has, for its size, gigantic electrostatic properties? And isn't it a marvelous coincidence that charcoal has an almost singular affinity for poisons? When noxious chemicals come in contact with the charcoal particle, they are sucked up into its myriads of little black holes with such force that they are, for all practical purposes, bound and tied. For example, lab experiments have shown that one quart of activated charcoal powder will adsorb eighty quarts of ammonia gas. Now that is a black hole! And we can actually see it!

But seeing does not always mean believing. Seeing what charcoal will do for some unpleasant stomach gas or minor insect bite, what it will do in a lab, what it will do to environmental pollutants, what it will do in a clinical setting, or what it will do for someone on some remote island does not really mean you have been convinced about its effectiveness as a remedy for sickness. Often, it isn't until providence thrusts us out into uncharted waters, without our usual support systems, and we are faced with a life or death situation, that we reach for the unseen and untried. Then it is that we have a new appreciation of "believing".

Poisoning

While doing development work in Nepal, just before the rainy season set in, the small clinic at which I worked hosted a "health camp". A group of American doctors, dentists, and their spouses made the exhausting six-hour hike up and down, and finally over into the Huwas valley. I can only describe the week as exhilarating and exhausting. Right from early morning, people began arriving until there were over a hundred thronging about on our small knoll. Each day more came. The clinic was a sea of people. Temporary stalls were set up out in the open for some of the doctors. The dentists and their student helpers worked together on the young and the old stretched out on makeshift tables under the sky. They were there from morning till dusk, and the lines hardly seemed to diminish. Huge painted canvases promoting the benefits of the "Eight Doctors" (Chapter 19) were erected, encircling half the area. The week disappeared into a flurry of helping and doing.

As some of the medical team rested up for the trek out, Joyce, the director's wife, led a number of the others on a walk down to the river. On the way, they passed a family having a picnic beside a small Hindu shrine. The translator mentioned that the family lived just below the clinic. After stopping for tea in one village, Joyce decided to hike home by a different route. As they passed by the local health-post/pharmacy, the health worker called them over. A small, four-year-old boy was lying there not responding to anything. The group recognized him as one of the children at the family picnic. Joyce, herself a nurse, tells what happened:

> "Among the group was a pediatrician. He examined the child and mentioned meningitis and a couple other serious possibilities as a cause for the child's condition. Not able to confirm a diagnosis, the doctor decided to give a large dose of antibiotic by injection. It would take some time for the antibiotic to take effect, so we stood around observing the child, and conversing. At some point, I described a similar case, but it had been from poisoning. The baby had not been treated and had died. I suggested giving charcoal. After hearing my story, the physician agreed that this child could also be suffering from poisoning, 'but poisoning from what?' he wondered. Nevertheless, he decided that charcoal was worth a try.
>
> "So I went to gather some coals from the nearest cook fires. We pulverized them as best we could and mixed the gritty powder in a four-ounce glass with water. We strained the mixture through a cloth and, because the child could not swallow, administered it through a small tube. We were able to get some down, and the child began to struggle against it. That encouraged us to keep trying, and eventually we were able to get about two ounces down. It was difficult to get it into the child because it was gritty and the tube was too small. As the others worked, another woman and I quietly offered a prayer that God would add His blessing to our efforts.
>
> "Very soon the child's breathing, which had been shallow and irregular, returned to normal. We removed the tube, and before the child totally refused to take anymore, we were able to get one more ounce down. By

> then the boy was completely alert. From the time we were able to get the first bit of charcoal down to the time he was back up and running around was no more than five minutes. We were all absolutely amazed! The doctor insisted that it had to have been the charcoal, because the antibiotics could not possibly have worked so quickly."

I hope those of you who are still skeptical never have to face such choices. But, by the time you've finished these pages and listened to those who have, I hope you will have enough evidence to try charcoal yourselves should an emergency arise. Faith is very powerful, but we need to add knowledge to our faith.

Charcoal

Charcoal is not classified as a drug or as a mineral—minerals are defined as inorganic. You may see it sometimes listed as a food supplement but this is quite incorrect, since it is both completely inert and indigestible as far as humans are concerned. Charcoal is unique in that there are really no other elements or compounds with which it can be grouped.

So, what exactly is charcoal? Put simply, as wood burns there is often not enough oxygen to allow for complete combustion. The water evaporates off, and the carbon in the wood distills into the black charred coals or crust we see when we put the fire out. For thousands of years, men have refined this process and commercially fabricated charcoal to be used as a smokeless concentrated fuel. Large mounds of wood, typically hardwoods, are arranged in such a manner that, once the fire is ignited underneath, the wood smolders for a period of time, slowly drying and eventually changing into charcoal.

Brief History

Because it burns hotter, charcoal is superior to wood, and so, historically, it became the fuel used to smelt ores. 1750 B.C. is its earliest known recorded use. The Egyptians and Sumerians produced charcoal for the reduction of copper, zinc, and tin ores in the manufacture of bronze. This practice continued through the Bronze and Iron ages into the early 18th century, when charcoal was replaced by coke made from coal. Today, the metal industries use activated charcoal for purifying brass and copper or for hardening steel. Charcoal was also the fuel of choice in the history of glass making. In this 21st century, it is now a vital ingredient in fuel cells designed to power the newest generation of motor vehicles.

Today, whether for preparing supper on an uptown balcony, in a reputable restaurant, or in a primitive hut, charcoal is still used to cook those favorite savory dishes. But, by the 20th century, charcoal's reputation and versatility had begun to spread far beyond providing a clean burning fuel. Methods were developed to make 'activated' charcoal, which introduced to the world the real genius behind the lowly charcoal.

Some Facts

Distinct from its heating qualities, the wonder of charcoal comes from its ability to "**ad**sorb"—to cause the atoms or molecules of a substance to form on its surface. This function is not to be confused with "**ab**sorb", as when a towel soaks up water. The tiny particles of charcoal are riddled with a network of crevices, cracks, and tunnels such that the combined surface area in a one-centimeter cube unfolds to a thousand square meters! This tremendous surface area, together with its capillary action, electrostatic properties, and other undefined properties, make it the indisputable champion of detoxifiers. Whether gases, foreign proteins, body wastes, free radicals, poisonous chemicals or drugs, charcoal binds the offending agents until they can be safely disposed of.

Activated charcoal begins as regular charcoal and is then "activated" with oxidizing gases, such as steam or air, at high temperatures. This oxidative process further erodes the charcoal's internal surfaces. This increases its adsorption capacity by creating an internal network of even smaller pores rendering it two to three times as effective as regular charcoal. But charcoal is not produced from wood alone. Bone char, coconut shells, peat, coal, petroleum coke, and sawdust are the most common starting materials for making activated charcoal. Many other materials have been experimented with, but generally are not as economical.

Charcoal Products

Charcoal is divided into two major commercial product groups. There is the larger granular activated charcoal (0.5-4 mm range) and powdered activated charcoal with a very small particle size (1-150 micron). This translates into an internal surface area anywhere from 500 up to 1500 m^2/gram. This vast internal network of cages is accessed by way of pores that also range in size from very small to even smaller (from over 50 nm in diameter down to under 2 nm). By knowing what is to be adsorbed, science and technology are able to design specific activated charcoals with just the desired pore sizes. By combining the right raw materials (each have their unique qualities) and activation conditions, one manufacturer is able to offer up to 150 grades of activated charcoal, each suited to a certain job.

Manufacturers of charcoal first look at the impurities to be removed from a substance. They then select the material and activation process that best match the impurities. For example, proteins are generally large molecules and are found in most natural products. Next come coloring compounds. The range continues all the way down to the very small, relatively volatile odor compounds. Industry has been able to produce charcoals with internal structures and pores that match these various molecular sizes.

While targeting specific impurities, activated charcoal works through several different mechanisms. Adsorption, the most well known mechanism, acts by physically binding molecules to the charcoal—technically known as Van der Waals forces, or chemisorption. Then there are chemical reduction reactions—as when charcoal is used to remove chlorine from water. Activated charcoal can also catalyze a number of chemical conversions, or can be a carrier of catalytic agents such as precious metals. An example is using silver impregnated charcoal to disinfect water. Charcoal can also act as a carrier of biomass, as in supporting material in biological filters used in your backyard goldfish pond. Another function is as a carrier of chemicals as in slow release color applications—food dyes and pigments. Are there more mechanisms? Scientists suspect there are.

Buckyballs

What makes charcoal different from other carbons, like the soft graphites used in pencils or those found in hard diamonds? It has long been known that the carbon rings in graphites lie in planes and are easily shed one from another, while the structures in charcoal are more like a spherical latticework.[1] Also, unlike the impregnable structure of diamonds, charcoal offers access to its interior. But there is more to charcoal than that.

In 1985 researchers H.W. Kroto and R.E. Smalley were curious about the atmosphere of giant red stars. It was known that carbon forms cluster molecules under such conditions. Among other carbon species, they detected the carbon molecule C_{60} for the first time.[2] It possessed unique physicochemical properties, extra stability, as well as some previously unexplained phenomena. To account for these features, they proposed a geodesic-like structure, one that essentially looks like the pattern on a soccer ball. Consequently the molecule was named after Buckminster Fuller, the inventor of geodesic domes (made famous at the 1967 World's Fair). Buckminsterfullerene (fondly referred to as "Buckyballs" amongst some researchers) is the chosen name for C_{60}, whereas the name fullerene is conveniently used for this whole family of closed carbon cages. They may not be as big as giant red stars, but these microscopic cells are just begging to be filled.

In 1999 Eiji Osawa, and colleagues at the Toyohashi University of Technology in Japan, demonstrated that C_{60} can also be extracted from wood charcoal.[3] As a result, many researchers now visualize charcoal as a structure made up of fragments of these "Buckyballs". Along with the discovery of nanotubes or "Bucky onions" there is the suggestion of new magnetic and electrical properties. It all sounds a little bit like science fiction. No doubt in time these latest models for charcoal will again be modified. In the meantime charcoal still mystifies even the informed. Scientists marvel as they continue to ask, "How is charcoal able to ...?"

Henry Schaefer is the resident Quantum Chemist at the University of Georgia and five-time nominee for the Nobel Prize. He is the third most quoted chemist in the world. He writes of his different discoveries: "The significance and joy in my science comes in those occasional moments of discovering something new and saying, "So that's how God did it!"" How is charcoal able to ...? God knows.

A Mystery

Being skeptical by nature, I want to know how a thing works. Naturally then, I want to know why/how charcoal works. I can hardly blame some people for questioning charcoal as a remedy. If they ask how it works, science has no complete answer. For some minds, that will mark the end of their interest. For those who may be critical because there is no complete "rational" explanation as to how charcoal works, may I point you to the classic drug reference for students and practitioners, *The Pharmacological Basis Of Therapeutics* by Goodman and Gilman. Goodman writes, "There are few drugs, if any, for which we know the basic mechanism of action. Drug action is not drug effect. The effect results from the action of the drug." Regarding the use of nitroglycerin to moderate attacks of angina, Gilman goes on to write, "The mode of action of nitrates to relieve typical angina is not fully understood." The same may be said of charcoal. The positive effects of charcoal as a remedy are well documented, as you will see in the following chapters. While the question of how charcoal works remains unclear, as it does for most drugs, there is one big difference. Charcoal has no known poisonous side effects.

From the dawn of civilization, man has had an intimate relationship with charcoal. As man and his technology have advanced, so too have the environmental consequences increased. History tells us vast tracts of forests were denuded to meet the demand for the prized charcoal to change rock and sand into metals and glass. But the metals and glass used to build, beautify, and conquer with were a two-edged sword. As the land suffered from the thoughtless clear-cutting practices, the atmosphere and streams also began to feel the added burden of industry's pollution. And, with the polluting came new diseases.

As time has marched on, technology has proliferated and, not surprisingly, so too has disease, on land, in air and sea, in plants, in animals, and in man. Two thousand years ago, in the midst of the Iron Age, Paul of Tarsus recognized even then that, "The whole creation groans and labors as in the pains of childbirth right up to the present."[4] How well he described our environmental concerns today.

But, we have not been abandoned to our wastelands of pollution and disease without help. The simple benefits of the humble charcoal have also been magnified and exploited, almost beyond imagination. Charcoal has been found more than capable of disarming the lethal properties of thousands of man-made pollutants. Science is employing charcoal in countless ways as it tries to stem the tide of exponential growth in disease in both the animal and vegetable kingdoms. Whether you are living within easy access of the technological advances of North America, or hiking the trails of some developing country, charcoal is just as modern as it was four thousand years ago, just as universal, just as versatile, and just as powerful. In a world being poisoned by its own near-sighted wisdom, God the Creator has provided man with a microscopic black hole big enough to swallow much of what ails us. CR

---

[1] Krätschmer W, Lamb L D, Fostiropoulos K, and Huffman D R. Solid C_{60}: a new form of carbon. *Nature* 347, 354–358, 1990

[2] Kroto, H W, Heath, J R, O'Brian, S C, Curl, R F and Smalley, R E C_{60}: Buckminsterfullerene. *Nature* 318, 162–163, 1985

[3] M Shibuya, M Kato, M Ozawa, PH Fang and E Osawa, Detection of buckminsterfullerene in usual soots and commercial charcoals, *Fullerene Science and Technology*, 7, 181-193, 1999

[4] Bible, Romans 8:22

Chapter 5 Submarines, Space Stations, and Roses

A walk through the woods can give you a first clue as to charcoal's amazing preservative qualities. For several years I lived in southern British Columbia, Canada. Often, while hiking up into the mountains, I would come across the remains of giant fir and cedar trees. There they stood, seemingly defying time and the elements. But almost always their trunks were still blackened with patches of charcoal. In past decades, they had apparently survived a forest fire. Yet there they stood, somehow preserved from the fate of much younger trees lying all about in varying stages of decay and rot. While nearby living trees were covered with moss or fungi, the charred trees stood bare.

Refreshed by my invigorating outing, I would return to my little mountain refuge. If I wandered into my huge old barn, I might once more ponder why the logs supporting the second floor were all scorched black. One day it hit me. Of course, the original homesteader had either collected snags once scorched by a forest fire, or he had purposely singed them himself. No doubt he was simply following an age-old tradition that dates back to ancient Egyptian times. Charcoal posts were used for construction support in wet areas where ordinary timbers would have quickly rotted.

Water Treatment

Centuries later, wood tars produced from charcoal were used for caulking ships. Recent studies of the wrecks of Phoenician trading ships from around 450 B.C. suggest that drinking water was stored in charred wooden barrels. This practice was still in use in the 18th Century for extending the use of potable water on long sea voyages.[1] Wood-staved barrels were scorched to preserve them, and the water or other items stored in them. How ingenious it was, a completely natural, organic, and environmentally friendly preservative!

Hindu documents, circa 450 B.C., refer to the use of sand and charcoal filters for the purification of drinking water.[2] One historical researcher believes the Bible too carries veiled references to the known cleansing properties of charcoal. Some see in the book of Numbers (19:9) and elsewhere the ancient use of charcoal for water purification.

Here is a note in the 1881 book of general knowledge, *The Household Cyclopedia*, under "Purification of Water by Charcoal":

> "Nothing has been found so effectual for preserving water sweet at sea, during long voyages, as charring the insides of the casks well before they are filled. Care ought at the same time to be taken that the casks should never be filled with sea-water, as sometimes happens, in order to save the trouble of shifting the ballast, because this tends to hasten the corruption of the fresh water afterwards put into them. When the water becomes impure and offensive at sea, from ignorance of the preservative effect produced on it by charring the casks previous to their being filled, it may be rendered perfectly sweet by putting a little fresh charcoal powder into each cask before it is tapped, or by filtering it through fresh-burnt and coarsely powdered charcoal."

Today, in its activated form, charcoal is used to filter water for small fish aquariums right up to Olympic size pools. It is a vital component of domestic and commercial water systems, as well as recreational water filters. Are there other uses?

The Many Faces of Charcoal

Its reputation for purifying water is historical, but charcoal's ancestry goes back even further. What child has not grabbed a burnt stick from a campfire and explored his artistic talent on some natural canvas? In fact charcoal art predates any of its other numerous uses. Even to this day, artists attract crowds in shopping malls as they sketch faces and landscapes. Bamboo charcoal is the principal tool in Japanese Sumi-e art. Charcoal has kept pace with man's history as a common ingredient (charcoal black) in inks and dyes. During the 18th century the European version of India ink was produced by combining lamp black from the soot of a variety of sources including charcoal. How many things can claim to have recorded their own history for thousands of years?

In the 12th century A.D., the Chinese included charcoal in black powder for their fireworks. This graduated to gunpowder for weapons and explosives. In our day, charcoal technology has literally exploded into hundreds of unrelated products. It is used in filters in industrial as well as military gas masks. Nuclear submarines would be useless without charcoal to scrub the air of carbon dioxide. The food industry uses it to clean obnoxious odors or flavors, and unwanted pigments from different foods, including our cooking oils. On the other hand, activated charcoal's function in licorice and some jellybeans is to add color. The pharmaceutical companies are absolutely dependant on charcoal in cleaning and concentrating their drugs. Charcoal made from animal bones is used to whiten table sugar. Photo and chemical labs use it layered into fabric blanketing the walls to adsorb lethal air-born poisons. With the growth of environmental agencies, charcoal has become a principle resource used in toxic spills and clean-ups. Because of its diversity of uses, one manufacturer boldly claims, "Virtually every product manufactured today has been improved upon at least once by the use of activated carbon."

What other fundamental uses for charcoal have we missed? There is its connection to the silk industry. During the first half of the twentieth century the main use for charcoal was for the production of carbon disulphide—a chemical used in making artificial silk. As well, silkworm oil is extracted from silk cocoons with petroleum ether, then refined by neutralizing it with alkali and bleaching it with activated charcoal. The cosmetic industry uses the charcoal-filtered oil in skin and hair moisturizing and conditioning products. Who would have guessed?

Back at my mountain retreat, my boys and I would sometimes go out along the Salmo River to pan for gold. As I thought about gold extracted on a large industrial scale, I wondered how the chemicals, used to separate the gold from other heavy sediments, were removed. You guessed it, charcoal is also utilized in gold recovery.

Charcoal is used as a top dressing for gardens, lawns and greens. Used in potting soils and bedding compounds, charcoal works as a soil sweetener while it neutralizes pesticides and herbicides. It is both a fertilizer and an insecticide for roses. Charcoal "by any other name would be as sweet".

The list is almost as universal as it is for water and no more dangerous. A word search for "charcoal" on the Internet will take you far beyond this short chapter, even into outer space. Yes, NASA uses charcoal in both its water and air purification systems onboard the International Space Station. With the renewed threats of chemical and biological terrorism, charcoal's use in neutralizing mustard gas and other chemical warfare agents is being researched.

As mentioned, the more common industrial sources of charcoal today come from petroleum coke, coals, peat, sawdust, wood char, paper mill wastes, bone, and coconut shells. It is an odorless, tasteless powder. Charcoal from vegetable sources such as wood or coal has about 90% carbon, whereas bone charcoal contains about 11%. The rest are calcium salts. Since it readily adsorbs impurities from the atmosphere, especially in powder form, but also in capsules and tablets, charcoal should be stored in a tightly sealed container. And speaking of adsorbing impurities from the air, there are the HEPA air filters used in industrial settings. You, no doubt, have one right in your own home. They are the industry standard and were developed for the Atomic Energy Agency.

Going back again in time, there was an early Egyptian practice of wrapping the dead in cloth. They were then buried in charcoal and sand to preserve the corpses.

This was later improved upon by collecting byproducts of charcoal for use in their embalming industry.

In review, we drink water filtered by it; breath air scrubbed with it; eat food purified through it, wear clothes made with it; grow our food and flowers in it; go to war with it; preserve things in it; enjoy hundreds of dishes cooked by it; we move mountains with it; we make the night sky sparkle with it; we take it with us to the bottom of the deepest oceans and out into space; swim in water washed with it; paint or draw our inspirations with it; and we record man's history of successes and mistakes dipped in it. Not least and not last, charcoal is called upon to clean up many of our technological mistakes. No wonder we naturally warm up to it.

Now, if charcoal is more than able to single-handedly take on such a host of dangerous elements; if it is so well suited to such a variety of tasks not only in the industrial/commercial world, but also in our very homes; and if charcoal can accommodate different techniques of delivery; don't you think it might have a significant role in maintaining, restoring and enhancing man's level of health? If charcoal's antifungal and antibacterial properties preserve dead trees and dead humans, how about living things? Stay tuned. *CR*

---

[1] Carbon Materials Research Group, Historical Production and Use of Carbon Materials, Center for Applied Energy Research, University of Kentucky, December 5, 2003

[2] *Ibid.*

Chapter 6 A Lamb Among Dragons

For those of you who, like me, enjoy the richness of history, included here are an assortment of experiences from the late 19th and early 20th centuries. This was a time in the development of charcoal, when its application as a healing remedy was being widely experimented with. It was also a time when doctors typically resorted to extremely poisonous concoctions. By comparison, charcoal was a lamb among dragons.

In Europe, interest centered mostly on charcoal's antidotal properties with respect to poisons. Whereas in America, physicians were exploring other possibilities for charcoal, such as intestinal disorders. As an example, here are two advertisements familiar to that time. In 1908 the Sears, Roebuck and Co. catalogue carried this:

> **Willow Charcoal Tablets** — "Every person is well acquainted with the great benefit derived from willow charcoal in gastric and intestinal disorder, indigestion, dyspepsia, heartburn, sour or acid stomach, gas upon the stomach, constant belching, fetid breath, all gaseous complications and for the removal of the offensive odor from the breath after smoking."

Here is another ad extolling charcoal's antibacterial and antiparasitic qualities:

> **Bragg's Vegetable Charcoal and Charcoal Biscuits** — "Absorb all impurities in the stomach and bowels. Give a healthy tone to the whole system, effectually warding off cholera, smallpox, typhoid, and all malignant fevers. Invaluable for indigestion, flatulence, etc., eradicate worms in children. Sweeten the breath."

Are these just wild claims of the day? As you will see in the following chapters, these and other claims have since been repeatedly substantiated by modern scientific research.

Of the different schools of medical practice in the late 1800s and early 1900s, it can be shown that, despite all their different strategies for treating disease, doctors at least agreed on the efficacy of charcoal. Let's begin with a few of the medical textbooks of the day. It should be noted that, when 'animal' charcoal is mentioned, it is referring to charcoal produced from animal bones. The reader is left to research for himself the more archaic medical terms. The portions described approximate modern measurements.

King's American Dispensatory of 1898

The authors Harvey Wickes Felter, MD, and John Uri Lloyd, Phr M, PhD, were both professors at the Eclectic Medical Institute, Cincinnati, Ohio. The many different therapies discussed in the book were carried out in both clinical and hospital settings. This is a revision of an earlier work by John King, an encyclopedic text that encompasses the entire materia medica of the Eclectic physicians of the 19th century. It covers botany, history, chemistry, uses and dosage. It is still referenced today by serious practitioners of botanical therapeutics. The Eclectic school of medicine dates back to the 1840s as part of an immense anti-medical reform movement in North America. It was a rural, primary care directed practice of medicine. They discarded the more poisonous drugs and drastic forms of treatment, and emphasized concentrated herbal medicines:

> "It [animal charcoal] has likewise been highly extolled as an internal remedy, in doses of a half grain to three grains, twice a day, in *scrofulous* [a tuberculous condition of the lymph glands] and *cancerous affections, goitre, obstinate chronic glandular indurations* [tumors], etc. ... Like vegetable charcoal, it destroys the odor of putrid animal matter. Dr. A. B. Garrod, in a paper read before the Medical Society of London, Nov. 17, 1846, recommended purified animal charcoal in cases of *poisoning* by opium, strychnine, aconite, belladonna, stramonium, tobacco, hemlock, etc."[1]

In the same book, under *Carbo ligni* (wood charcoal), we read of its varied uses when taken internally:

> "It acts as an absorbent (both fluids and gases) and disinfectant. Its internal employment will be found useful in those *digestive derangements* which are associated with offensive breath and disagreeable belchings; also to correct the fetid condition of the stools in *dysentery*. It is also useful in *acidity of the stomach, flatulency* [gas], and in the nausea and constipation attending *pregnancy*. It is also very useful in internal heat and irritation of the stomach, with acidity; *sick headache; diarrhoea; cholera infantum,* etc. In cases of sick *headache*, due to gastric acidity or derangement, and which are ushered in with blurred vision, photopsia, and finally nausea and intense headache, I have found a drachm [dram] of charcoal mixed in a little syrup, to which is then added about a gill of water, and ten or twelve drops of ether, to afford prompt relief; in very obstinate cases, the dose may require to be repeated two or three times, every twenty or thirty minutes (J. King).

In some cases charcoal may be advantageously combined with the subnitrate of bismuth as a sedative; and where a laxative action is required, rhubarb may be beneficially added to it. Bilious colic is said to have been cured by it, in doses of one drachm in two fluid ounces of burnt brandy, repeated as required.

"Externally, it may be used in poultices to correct fetor of *ulcers,* arrest *gangrene,* etc., and is efficient in many *cutaneous* [skin] *diseases*. It absorbs foul gases generated in vaults, sewers, etc. It is also a useful haemostatic [agent to control bleeding], having arrested *epitasis* [nose bleeding] when subsulphate of iron had failed.

"The *specific* use of charcoal," [says Dr. Scudder (*Spec. Medication*)] "is to arrest *hemorrhage from the bowels*. It has been used in enema, finely powdered, to four ounces of water, thrown up the rectum. Why this checks it I can not tell; that it does it, I have the evidence of my own eyes. For several years I have employed the second decimal trituration [finely ground powder] as a remedy for *passive hemorrhage,* with most marked benefit. I employ it in threatened hemorrhage during *typhoid fever*; in *menorrhagia* [abnormally high menstrual bleeding], especially when chronic; in *prolonged menstruation*; the watery discharge that sometimes follows menstruation; *hemorrhage from the kidneys*; *hemorrhage from the lungs*; and in some cases of *leucocythemia*. A good indication for this remedy is a small, pallid tongue with lenticular spots, and with this it may be given in any form of disease." It occasionally enters into tooth-powders, and may be used with advantage to correct the fetor of the mouth, and cleanse the teeth."[2]

Here is one more unusual entry under *Fucus* (a common seaweed like kelp):

"On account of the iodine contained in its charcoal, known as Vegetable ethiops, it has been found beneficial in *scrofulous enlargements* of the [neck] glands; the plant being incinerated in a covered crucible, and the charcoal given in doses of from ten grains to two drachms."[3]

Apparently, the charcoal captured the iodine from the kelp when it was burned and then liberated the iodine in the digestive track when the charcoal was taken internally. This proved beneficial in the treatment of goiter.

Under Charcoal *cataplasm* (poultice) we read:

"*Preparation*: Macerate bread, two ounces, with water, ten fluid ounces, for a short time near the fire; then gradually add and mix with it powdered flaxseed, ten drachms, stirring so as to make a soft cataplasm. With this

> mix powdered charcoal two drachms, and when prepared for application, sprinkle one drachm of charcoal on the surface of the cataplasm. *Action and Medical Uses.* Charcoal, properly prepared, has the property of removing the fetid odor evolved by *gangrenous* and *phagedenic* [rapidly spreading] *ulcers*, for which the above cataplasm is designed. It should be renewed two or three times in every twenty-four hours."[4]

Clearly, charcoal was not only recognized as a valuable remedy, but was used internally and externally as a poultice for a variety of ailments. You will find that the combination of flax seed and charcoal used in poultices is still popular today. The flax seed, when powdered or boiled whole, acts primarily as a binder for the charcoal which, by itself, dries out fairly quickly. Flax seed is also known for its own healing virtues.

The Eclectic Materia Medica, Pharmacology & Therapeutics, 1922

This book was also authored by Harvey Wickes Felter, MD, and one of the last Eclectic medical publications as the Allopathic school of medicine gained popularity. The entries were numerous and quite detailed. Some twenty-five years after his earlier book, we see charcoal is still recognized as a valuable remedy among some doctors:

> "*External.* Absorbent, deodorant and disinfectant, but not antiseptic. It is used very largely to deodorize foul ulcers, carcinomata, and gangrene, possessing the advantage of being an odorless deodorant. It is frequently added to poultices and is an ingredient of some tooth powders. A rectal injection of charcoal has checked hemorrhage from the bowels …
>
> "*Internal.* Its absorbent and deodorant properties make charcoal a splendid agent to absorb putrid gases from the stomach and bowels. It is indicated by offensive breath and disagreeable belching. In acidity of the stomach, gastric distention, nausea and vomiting, sick headache with gaseous belching, fetid diarrhea, and sometimes in the acid vomiting of pregnancy, charcoal is a most effective agent. It may be combined, plain or aromatized with oil of peppermint …"[5]

The Physiomedical Dispensatory, 1869

This is another eclectic commentary of North American herbalism and focuses on mild herbs, tonics, long-term herbs, and non-drastic measures. After explaining its tremendous ability to adsorb gases, author William Cook, MD, describes charcoal this way:

> "This charcoal is used internally to relieve flatus in the stomach and bowels, following indigestion. It does not strengthen the digestive apparatus, but merely affords ease from present mechanical distention. For this purpose, from five to ten grains may be given an hour or more after a meal. It does not interfere with the action of suitable tonics; but partially weakens the power of the gastric juice, if given before the digestive process is completed. In bilious diarrhea with frothy stools; in all cases of fetid stools; and in acrid discharges which create tormina [acute, colicky pains], it is also of service. Among cases of the last kind may be named the accumulation of flatus in the last stages of peritoneal inflammation, when the distention of the bowel by gases will cause that 'angulation' which prevents stools and causes much suffering. In such cases, from three to five grains may be repeated every two hours; ... Charcoal is an antiseptic, whether used internally or externally. For this purpose, it is given in semi-gangrenous conditions of the stomach, and applied to phagedrenic [rapidly spreading] and gangrenous ulcers. It arrests the process of decomposition; but does not aid in preserving the deeper tissues, nor in building up a line of demarcation, nor in securing a granulating surface."[6]

The reader may have noticed the disagreement about charcoal's antiseptic quality. I am unclear as to why in one place charcoal is said to be antiseptic, and in another said not to be. In the different older publications I reviewed, charcoal is mostly described as antiseptic.

***The British Pharmaceutical Codex*, 1911**

This was the British equivalent of the *American Dispensatory*, for the use of medical practitioners and pharmacists. It includes hundreds of plants, some of them toxic, most not, and a few more or less toxic chemicals. It focuses more on single ingredients such as alkaloids and less on whole plants. It was written by pharmacists for pharmacists.

Here are a few interesting entries published by direction of the Council of the Pharmaceutical Society of Great Britain, 1911:

> "Action and Uses ... It is used internally as an antiseptic and absorbent, in flatulent dyspepsia [acid indigestion], intestinal distension, diarrhea, and dysentery. Its action is mainly mechanical, removing mucus and stimulating the movements of the stomach and intestine. Externally, charcoal is absorbent and deodorant. It is sometimes employed as a poultice for fetid ulcers, some of the charcoal being spread on the surface of the poultice to retain its oxidizing properties. The powder ... may also be administered

on buttered bread in the form of sandwiches. Lozenges of charcoal ... and charcoal biscuits are a popular form of administration. Charcoal tooth powders may contain from 25 to 75 percent of wood charcoal."[7]

They also provided this recipe for a Slippery Elm poultice:

Slippery Elm, in fine powder	3 ounces
Boric Acid, in powder	½ ounce
Wood Charcoal, in powder	½ ounce
Boiling Water	8½ fl. ounces

Mix the powders and add the boiling water gradually.

It was said to be a useful application for indolent ulcers and whitlows (a pus filled infection on the skin at the side of a fingernail or toenail). The Slippery Elm would give a jelly-like consistency but, by itself, it is used for burns, ulcers, chilblains, and skin diseases. I am unclear as to why the boric acid was included.

American Journal of Pharmacy, 1887

The Philadelphia College of Pharmacy, the oldest school of Pharmacy in the United States, founded in 1822, published the above monthly journal for over a century. It included this charcoal compound:

> "Charcoal and Camphor. A mixture of equal parts of camphor and animal (bone) charcoal is recommended by Barbocci for preventing the offensive odor and removing the pain of old excavated ulcers. The camphor is stated to act as a disinfectant, and the charcoal absorbs the offensive odors."[8]

While camphor is antiseptic in low dosages, it can also be quite irritating. It will be shown in the following chapters that charcoal by itself is usually adequate for controlling infection, as well as for eliminating foul odors. I would caution against actually using the above formula or any combinations of charcoal that include toxic substances, especially if taken internally. I should mention that in looking over old references, I also found charcoal occasionally being compounded with some very poisonous substances. As we will see in the next chapter, it was fairly well known during this period of its history, that charcoal easily neutralized the deadly effects of the most powerful drugs. Charcoal also adsorbs camphor. Since it was common knowledge that charcoal would adsorb at least a portion of the other lethal ingredients, one is left to wonder why the pharmacists who concocted some of these questionable cures included charcoal.

Eclectic Medical Journal, 1876.

John M. Scudder, MD, like scientists today, was repeatedly asked how different remedies worked. He addressed this perennial question in an article in the *Eclectic Medical Journal*. He uses charcoal as his first illustration. He writes:

> "*How* the remedies act to cure disease we do not know and can not know, and it is of far less importance to us than the simple fact that they will cure.
>
> "I could not give a plausible guess why a few doses of triturated charcoal, not a grain in all, should check a severe hemorrhage. And yet I know the fact as well as I know that the sun rose this morning. As an example, Thomas French, a man weighing over two hundred pounds, stout and full blooded, came to me complaining that he was having repeated hemorrhages from the nose that was rapidly exhausting him. It had been going on for some days, and the means employed had utterly failed. His face was pallid, the pulse soft and weak, extremities cold. I gave him ten grains of triturated charcoal to be taken in grain doses every three hours. There were three ineffectual efforts at hemorrhage after commencing the powders, but it was effectually stopped the second day! Now if this was but a single case we would think but little of it, but I have repeated it scores of times with the same result."[9]

Etiquette and Advice Manuals, 1893

Notice this womanly advice from Baroness Staffe, in a chapter entitled, *The Lady's Dressing Room*:

> "It is also advisable, when the feet are swollen from a long walk or much standing, to bathe them in water in which charcoal has been boiled. The water should be strained through a cloth before putting the feet into it. Swelling and fatigue will both disappear rapidly. Alcoholic friction is also very good."[10]

Blood circulation in our feet is more sluggish than in other parts of the body, and so swelling of the feet can be a problem. The foot bath with charcoal would help to stimulate circulation and to draw out toxic waste products from the blood that might pool in the feet. Considering the source of the remedy, we can see charcoal can be "refined" in more than one way.

The Household Cyclopedia of General Information, 1881

In this prodigious book of general knowledge, compiler Henry Hatshorne, MD, included this entry for *Charcoal Poultice*:

> "To half a pound of the common oatmeal cataplasm, add two ounces of fresh burnt charcoal finely pounded and sifted. Mix the whole well together, and apply it to foul ulcers and venereal sores; the fetid smell and unhealthy appearance of which it speedily destroys."

Here is another entry:

> "Charcoal is a useful deodorizer and purifier; it acts by its attraction for organic matter and gases. By condensing the latter as well as the oxygen of the air in its pores it causes rapid combination. Small animals buried in charcoal are rapidly converted into skeletons, while no offensive smell is noticed even in warm weather. Water is best kept in charred casks; foul water is purified by filtration through charcoal. Meat lightly tainted is restored by wrapping in powdered charcoal; animal [bone] charcoal is the best. Lampblack is nearly worthless for these purposes. Animal [bone] charcoal is an antidote to all animal and vegetable poisons; it rapidly removes organic coloring matters and also vegetable bitters from solution."[11]

Three Square Meals, circa 1880

This next excerpt was taken from "The Toilet", a chapter in the cookbook *Three Square Meals*. The word cookbook is a misnomer because it was much more than a cookbook. It was in essence a household bible—probably a much treasured wedding gift of its day. It included recipes, house cleaning tips, laundry and care of clothing, soap-making, items on invalid care, how to make sickroom remedies, medicine and tonics and dyes:

> "Health is one of the requisites to the making up of a fine complexion. A sickly plant commands our care, but not our admiration. So with the individual. A buoyant step and healthful glow on cheek and lip, are irresistible in their power over us. To possess these the greatest care should be taken. Plenty of nutritious food well cooked and at regular intervals. Exercise in the open air. Early hours for rest and sleep are all absolutely necessary. Avoid medicine of a drastic and debilitating nature, and in the spring, when circulation is clogged and digestion sluggish, take a tablespoonful of French charcoal mixed carefully in water or honey before meals for several days, following this each evening with a teaspoonful of

extract of dandelion; or take the same dose of charcoal at night, follow it with a large spoonful of finely minced onion. There is no greater purifier in the medical pharmacy than charcoal. In the spring of the year, eat freely of cabbage, lettuce and all herbaceous food. If this diet is accompanied and followed by the requisite amount of bathing, it will work wonders with the most stubborn complexion and give health and elasticity to the sluggish frame. If spring tonics are prescribed, never take them until after charcoal has been used as above directed, when the system will be found in a state to be benefited by their use… Powdered charcoal easily removes stains and makes the teeth white, though it occasionally works under the gums… Charcoal may be mixed with honey if it is used for a dentifrice."[12]

Ellingwood's Therapeutist, **1908**

Finley Ellingwood, MD, was a clinician who specialized in obstetrics and gynecology. He was of the American Eclectic school and began this journal as a forum for physicians. He includes this note about charcoal and its known properties as an antidote for poisons:

"It is stated by those who are apparently informed, that many Japanese physicians use charcoal as a general antidote to agents which have been taken into the stomach which will produce poisoning.

"These observations, it is stated, have been farther proved by French physicians who claim that if powdered charcoal is taken soon after the ingestion of poison in very large quantities, its influence upon the poison will be noticed from the first.

"A tablespoonful may be mixed with a little water and taken frequently in divided doses, the whole amount within one or two hours."

The following was a leading article in the *Therapeutist* and was contributed by A.C. Hewett, MD, Chicago, Illinois:

"I recently received a 'hurry call' to attend Mrs. H. — 'very sick'. The patient, a woman in good circumstances aged about fifty years, was found to be in fact very ill; pale-gray in the face, forehead and limbs covered with a clammy perspiration, pulse so small and rapid that counting was next to impossible, suffering severe gastric and abdominal pain, 'had been vomiting copiously till nothing but a stringy mucus could be ejected'. Asked what she had eaten; I was told coffee, cross buns and canned boneless chicken; I at once diagnosed toxins, and gave charcoal, prepared as per the following Rx:

"Calcined willow charcoal and wheaten flour two heaping tablespoonfuls each; common table salt a level teaspoonful; warm water four ounces. The charcoal, flour and salt were first well mixed. Water was added little by little for convenience and speedy result.

"Dosage, a brimming tablespoonful every ten minutes regardless of recurrent vomitings. The first dose was partly ejected. The second retained in spite of attempts to vomit. After taking the third spoonful, pain and nausea gradually subsided. Of course a hot water bag and bottles were put to her feet, warmed flannels wrapped around her knees and a hot water bottle placed in her hands, which she smilingly and soon nested upon her stomach … I directed continuance of hourly doses of the charcoal mixture till all should be taken. An uneventful and rapid recovery ensued.

"Not long after I attended another patient similarly but not so severely affected. She was much younger; had "lunched," taking coffee, doughnuts and canned salmon. The charcoal mixture and applied heat, brought similar result, and the admonition to use charcoal in food poisoning."[13]

In spite of all the technical advances man has made, spoiled food, including spoiled canned food, and the many other health concerns of a hundred years ago, are still very relevant today. It seems clear that charcoal's medicinal value was well known and enthusiastically promoted by some doctors in past generations. It is a shame more doctors today are not aware of charcoal's proven track record when it comes to healing.

Certainly by the beginning of the 20th century charcoal had already established its credentials as an effective yet simple and economical remedy for dozens of health problems. But who could have guessed how much world fame awaited it. CR

---

[1] Harvey Wickes Felter, MD, and John Uri Lloyd, Phr M, PhD, *Animal Charcoal*, King's American Dispensatory, Ohio Valley Co., Cincinnati, 1898

[2] Ibid. *Charcoal*

[3] Ibid. *Fucus*

[4] Ibid. *Charcoal cataplasm*

[5] Harvey Wickes Felter, MD, Carbo Ligni, The Eclectic Materia Medica, *Pharmacology and Therapeutics*, 1922

[6] William Cook, MD, *Carbon, Charcoal*, The Physiomedical Dispensatory, 1869

[7] The British Pharmaceutical Codex, The Pharmaceutical Press London, 1911

[8] John M. Maisch, MD, editor, *American Journal of Pharmacy*, Vol. 59, 1887

[9] John M. Scudder, MD, *Eclectic Medical Journal*, 1876

[10] Baroness Staffe, *Etiquette and Advice Manuals - The Lady's Dressing Room*, (trans. by Lady Colin Campbell), *Victorian London Publications*, 1893 Part II www.victorianlondon.org/publications/ladys-2-2.htm

[11] www.mspong.org/cyclopedia/chemical.html

[12] http://freepages.genealogy.rootsweb.com/~twigs2000/toilet.html

[13] *Ellingwood's Therapeutist* - Vol. 2 #4 April 15, 1908

Chapter 7 Ancient and Modern Medicine

It was 1831 when, in front of a large group of his peers at the French Academy of Medicine, Pierre Touéry, a French pharmacist, reportedly drank a glass of deadly strychnine and survived to publish his story. There were no uncontrolled convulsions, no ill effects at all! Why? Because, he had combined fifteen grams of the poison (ten times the lethal dose) with an equal amount of charcoal.

In 1813, French chemist Michel Bertrand swallowed five grams of arsenic trioxide (150 times the amount that would have killed most people) mixed with charcoal. Again, there was no nausea, no vomiting, no diarrhea, no excruciating cramping, no severe burning in the mouth and throat, no collapse, no death. In a dangerous, but dramatic way, he had avoided the sure consequences of ingesting the arsenic and demonstrated charcoal's phenomenal ability to hold poisons from being absorbed by the body.[1] **But don't you try this at home!**

Obviously these men did not carelessly endanger their lives. By carefully observing laboratory animals, they knew how powerful and fast charcoal worked to neutralize poisons. They also knew some of charcoal's amazing history.

1500 B.C.

The first recorded use of charcoal for medicinal purposes comes from Egyptian papyri around 1500 B.C. The principal use appears to have been to adsorb the unpleasant odors from putrefying wounds and from within the intestinal tract. Hippocrates (circa 400 B.C.), and then Pliny (A.D. 50), recorded the use of charcoal for treating a wide range of complaints including epilepsy, chlorosis (a severe form of iron-deficiency anemia), vertigo, and anthrax. Pliny writes in his epoch work *Natural History (Vol. 36)*: "It is only when ignited and quenched that charcoal itself acquires its characteristic powers, and only when it seems to have perished that it becomes endowed with greater virtue." What Pliny observed and noted so long ago is the very mystery science continues to exploit today.

In the second century A.D. Claudius Galen was the most famous doctor of the Roman Empire, and the ancient world's strongest supporter of experimentation for scientific discovery. He produced nearly 500 medical treatises, many of them referring to the use of charcoals of both vegetable and animal origin, for the treatment of a wide range of diseases.

A.D. 1700

After the suppression of the sciences, first by Rome around A.D. 300 and then on through the Dark Ages, charcoal reemerged in the 1700s as a prescription for various conditions. Charcoal was often prescribed for bilious problems (excessive bile excretion). The use of charred wood was mentioned for the control of odors from gangrenous ulcers. It was also about this time that the sugar industry successfully developed a process using charcoal for the decolorization of sugar syrups. After the development of the charcoal activation process (1870 to 1920), many reports appeared in medical journals about activated charcoal as an antidote for poisons and as a cure for intestinal disorders.

A.D. 2000

Today, charcoal is rated Category 1, "safe and effective", by the U.S. Food and Drug Administration (FDA) for acute toxic poisoning. It is also listed in the U.S. homeopathic pharmacopoeia as having "marked absorptive power of gases". A 1981 study, reported in *Prevention* magazine, confirmed what Native Americans have known for hundreds of years. Activated charcoal cuts down on the amount of gas produced by beans and other gas-producing foods, and adsorbs the excess gas as well as the bacteria that form the gas.[2] Brand name, over-the-counter drugs may be more commonly used for gas because of their attractive packaging and commercial value, but they are certainly not as effective.

Modern research has validated most of the historic uses for charcoal, and has moved on to discover exciting new applications. Today charcoal is used in medical equipment that filters the blood of thousands of kidney and liver dialysis patients who might otherwise die. It is used by the pharmaceutical industry in the manufacture or purification of most medicines. It is found in emergency wards all around the world as the principal antidote for poisoning and drug overdose. These are but a few innovations in charcoal's growing portfolio.

Commercial Products

Commercial medicinal-grade charcoal, regular and activated, comes in five forms: tablet, capsule, paste, as a suspension in water, and powder. It can be found in many pharmacies and health food stores, and online on the Internet. If your local pharmacy or health food store does not stock it, they would most likely be glad to order it for you.

In America, you can be relatively sure of the high quality of the charcoal if it is listed as USP (U.S. Pharmacopoeia) grade. The same would be true for the national Pharmacopoeia grades in other countries. Commercial charcoal powder is most often the more adsorbent activated grade. Charcoal also comes in regular or vege-capsules. You will also want to note if the charcoal is compounded with other ingredients.

Charcoal tablets typically include fillers, binders and sweeteners. Research has not shown that the addition to charcoal of binders or sweeteners either increases or decreases the adsorption of poisons. In this book you will find references to charcoal being taken internally with other products to help with palatability, but the focus is on the purest form of charcoal.

When used as an antidote for poisoning, commercial charcoal suspensions often include sorbitol. However, besides being a sweetener, sorbitol also has a purgative or vomiting effect. In one study, all the individuals in the group who took a charcoal-sorbitol suspension experienced diarrhea, compared to none in the group who took a charcoal-only suspension. The addition of sorbitol did not improve the removal of toxins.[3]

There are other commercial products in which charcoal is compounded with very strong active ingredients. They are reminiscent of some concoctions of a hundred years ago and cannot be recommended in any amounts. These are not limited to homeopathic preparations. The fact that charcoal taken as a single ingredient is superior as an antidote for poisoning, remains unchallenged from both anecdotal experience and clinical evidence.

Commercial tablets, when compared to finely powdered charcoal, have been found to be about half as adsorptive. In one study with 182 individuals, pulverized charcoal was 73% effective in preventing absorption of a drug in the stomach, while tablets were only 48% effective. In most cases tablets are about 75% charcoal and 25% fillers and binders. Chewing the tablets well before swallowing will increase their effectiveness.

Charcoal will also help to sweeten the breath and with a few swishes of the toothbrush, it will help brighten your smile.

What Charcoal is Not

Before going any farther I need to dispel the myth that charred toast is a form of charcoal. The black crust is not carbon. It is composed of the very compounds that are normally consumed during the charcoal-making process. These include chemically changed super-heated proteins, fats and carbohydrates that may in fact contribute to cancer. I cannot prove or disprove that charred toast may be an effective home remedy, but it and other scorched foods are not charcoal and are not healthful. Since charcoal briquettes used for barbecuing are infused with poisonous petroleum distillates to help them ignite, they should never be used for medicinal applications.[4]

Charcoal is not a cure-all. However, its effectiveness in the treatment of different ailments has been documented in numerous medical journals, including the *Journal of the American Medical Association*, the *British Medical Journal*, *The Lancet*, and *Clinical Toxicology*. Today doctors, paramedics and medical centers use activated charcoal in a number of different ways: to eliminate toxic by-products that cause anemia in cancer patients; to disinfect and deodorize wounds; to filter toxins from the blood in liver and kidney diseases; to purify blood in transfusions; to cut down on odors for ileostomy and colostomy patients; to treat poisonings and overdoses of aspirin, Tylenol and other drugs; to treat some forms of dysentery, diarrhea, dyspepsia, and colic; to treat poisonous snake, spider and insect bites. More recently its ability to adsorb cholesterol has been reviewed.

Adverse Effects, Precautions and Contraindications

Charcoal will slow bowel movements in individuals who do not drink sufficient water, and charcoal may cause occasional stomach irritation in others with sensitive digestion, but these are not considered adverse side effects. It has also been suggested that charcoal may leave a tattoo if applied directly into a fresh cut or wound that may quickly heal over. I have never seen it happen or been told of it happening by others. But, if there is some concern, then simply apply charcoal inside a poultice. In some cases it may be necessary to weigh the slight possibility of a charcoal tattoo over the danger of contamination of a wound with toxins or bacteria. Charcoal is not necessary for an uninfected wound in an otherwise healthy individual.

Taken internally, charcoal will adsorb most medications thereby neutralizing any potential drug effect. For some this will be the very reason for taking charcoal. However, if one is taking a prescription drug, a safe rule of thumb is to take charcoal two hours before or after taking medications. This is the most common recommendation given by researchers and physicians. This does not rule out using charcoal in a bath or as a poultice in conjunction with oral medication.

The only published research I have found that lists charcoal as contraindicated is in the treatment of variegate porphyria (VP), a rare skin disease.[5] There are eight classes of porphyria. Initial research with congenital erythropoietic porphyria (CEP), or Gunther's disease, found activated charcoal to be helpful.[6] When later trials were made with VP, it was expected that there would be similar benefits with activated charcoal or, at worst, no effect at all. Instead there was a completely unexpected increase in skin disease, urine and plasma porphyrins. The results were said to be paradoxical and unexplainable. It may be that subsequent research will discover the cause of the out-of-character results of this first trial.

Nutrients

Charcoal is neither absorbed nor metabolized by the body.[7] However, some have wondered if charcoal binds essential food nutrients. Some research suggests that food in the digestive tract inhibits the effectiveness of charcoal. But, this is not interpreted to mean that charcoal is adsorbing food nutrients along with toxins. While some early studies seemed to indicate this to be true, several subsequent studies have not born this out. [8]

What we do know is that Russian researchers found that activated charcoal efficiently adsorbed toxins before these poisons could compete with oxygen and nutrients that were trying to pass through the cell membrane. It would seem then, that the very opposite may be true. Instead of adsorbing essential food elements, charcoal removes toxins that are competing with nutrients for intestinal and cellular absorption, thereby promoting efficient nutrient uptake. Those studies that suggest vitamins may adsorb to charcoal were done under abnormal laboratory conditions.

I would love to tell the reader categorically that charcoal does not depreciate the level of nutritive absorption in any way. While science has yet to prove this conclusively, it seems more prudent to say that if there is any adsorption of nutrients, it is so negligible that it has yet to be shown to compromise one's health. For instance, charcoal has been used for many years as a fecal deodorant for patients with ileostomies and colostomies. In spite of the fact that they may routinely take charcoal orally three times daily for years, it has never been demonstrated to nutritionally affect these individuals who are already at risk of nutritional deficiency.[9]

In one animal study, Dr. V.V. Frolkis, a famous Russian gerontologist, and his colleagues, demonstrated that the lifespan in laboratory rats increased up to 34% by feeding them charcoal in their diet![10] Toxins, including free radicals, are believed to play a significant role in aging. But these "loose canons" will form a stable matrix with charcoal in the gut until they are eliminated from the body. Researchers concluded that the binding up of these toxins in the intestinal tract before they are absorbed or reabsorbed into the system may be one mechanism that allowed the rats to live longer and healthier. Both clinical observation of patients in hospitals and numerous animal studies have demonstrated charcoal poses no threat to nutritional uptake.

Cholesterol

Will charcoal adsorb the nutrients from dietary supplements? Again, there is no conclusive research to say charcoal doesn't affect supplements, but we can safely say that if it does, the amounts are negligible. Having said that, cholesterol is an essential nutrient. But the body produces more than it requires and does not need to be supplemented with extra cholesterol from one's diet. In fact, for most North Americans the problem is having far too much dietary cholesterol, which, as we all know, increases our risk of death from coronary heart disease. This is where charcoal seems to be crafted to "think" for us to some degree, when we are not thinking for ourselves.

Charcoal lowers the concentration of total lipids, cholesterol, and triglycerides in the blood serum, liver, heart and brain. In one study on patients with high cholesterol, reported in 1986 in the British journal, *The Lancet*, two tablespoons (eight gms) of activated charcoal taken three times a day for four weeks, lowered total cholesterol 25%, lowered LDL cholesterol 41%, and doubled their HDL/LDL (high-density lipoprotein/low-density lipoprotein) cholesterol ratio.[11] Microscopic tissue examination shows that a daily dose of activated charcoal may prevent many cellular changes associated with aging—including decreased protein synthesis, lower RNA activity, organ fibrosis as well as sclerotic changes in the heart and coronary blood vessels. We can safely say, as an anti-aging adjunct to a total health program, the above cumulative effects of charcoal upon one's blood chemistry, may add up to a longer life and improved overall health.

The reason these astounding benefits of charcoal have not been more widely advertised will be considered in Chapter 21.

Food Quality

The food industry understands that as people become more knowledgeable about good nutrition, they also become more demanding in matters of food quality. More and more people want their food as natural as possible. What we also know about charcoal and food is what the food industry has known for a long time. Charcoal effectively and specifically removes those colorants and odors that are either unpleasant or unappealing from a host of foods, while enhancing the desirable flavors, without depreciating at all the nutritional content. Charcoal is sometimes used as a coloring agent in some foods. All these finishing touches come at no expense to food quality.

Detoxifier

Very few health practitioners realize that as an agent to remove toxins from the body, charcoal is the best single detoxifier for whole-body cleansing. Of course, along with charcoal, one cannot forget water, both on the outside and the inside to help wash away poisons.

Richard C. Kaufman, BS, MS, PhD, (Bio-nutritional Chemistry from the University of Brussels) has written extensively in the field of anti-aging. He writes, "Detoxification is an on-going biological process that prevents toxins (from infectious agents, food, air, water, and substances that contact the skin) from destroying health. Chronic exposure to toxins produces cellular damage, diverse diseases, allergic like reactions, compromised immunity and premature aging." As a general detox plan to counteract these daily exposures to toxins, he has found two programs that, using activated charcoal, have worked well for him and others. He recommends either.

1. Use activated charcoal on two consecutive days each week. Take a total of 20 to 35 grams each day divided into two or three doses. Take in the morning, at midday and before bed on an empty stomach. Avoid excessive calories or processed foods on those days.
2. Take about 20 grams a day of activated charcoal in divided doses for several months. Follow with a one-month break and resume the cycle.[12]

These programs should not be viewed as license to continue any unhealthy practices. Charcoal should be viewed as an auxiliary to sound health practices, not as a back-up for intentional indiscretions. But, should we give in to some carelessness, charcoal is a wonderful aid to recovery.

Puffy

One precaution that should not be ignored is that activated charcoal in its powder form can be … well, "explosive". Let me explain. I am not referring to using it in gunpowder or fireworks. Dry activated charcoal powder is so fine and light, that once you dip a spoon into it to put it into a glass of water to drink or to mix it into a poultice, the powder will easily float, forming a gray/black cloud. If you drop any amount of the powder, it tends to "explode" on impact shooting it everywhere. Accidents happen, but with time you will realize that the very nature of charcoal requires one to move guardedly. That is why most people prefer to take charcoal as a tablet or capsule.

But charcoal's reputation for spreading does not end by swallowing a capsule. A close friend shared this story on strict terms that, if I wanted to remain friends, I did not disclose any name or location. "MD" was preparing to open her place of work one morning when she felt the onset of an upset stomach. She stopped for a moment to take a charcoal capsule. She swallowed it down with some water, but it got stuck in her throat. She tried and tried to swallow it but it would not go down. She thought maybe if she tried a little burp that would dislodge it. To her great horror, when she burped, a black cloud of smoke exploded out of her mouth. MD said chuckling, "It frightened me at first, because I didn't know what it was. It also peppered the front of my blouse. I was mortified that someone might see me, so I hurriedly tried to dust myself off."

Charcoal does wipe up easily off hard surfaces, but once cloth is stained, it is permanently discolored.

Potential Benefits

I have refrained from trying to explain the mechanisms by which charcoal performs its wonders. Science is able to say, "This is what occurs" (the effect), but not how it happens (the mechanism). In doing the months of research for this book, I have come across conflicting claims as to how charcoal works. Certainly much more is known now than was known a hundred years ago. But I don't know how charcoal works, and I suspect that no one does completely. The interconnectedness of the human body is exceedingly complex, and the microscopic properties of charcoal are no less mysterious. One day someone may be able to explain the mechanisms perfectly, but until then, what one can say with certainty is that charcoal does work.

It's much like a car or computer. Most of us have very limited practical knowledge of the inner mechanics of these tools, but it doesn't prevent us from using them to our benefit. Clearly, they can be abused or used improperly, but with some basic knowledge they can serve us well. The same is true for charcoal. It will work whether one believes or doesn't believe, but it works even better when it is applied with some care and understanding of its mechanisms.

Remember the forest trees and that deadfall blackened by fire? There we saw charcoal's natural antibacterial and antifungal properties at work. Modern medicine continues to tap into that potential, as science tries to understand how charcoal makes it happen.

HIV-I and Hepatitis

The development of new virus inactivation procedures has become an area of growing interest. This is mainly due to increased demands concerning the safety of biological products, especially blood products and recombinant proteins used in medicine. Photochemical processes now represent the most promising methods to inactivate viruses. To date, dyes have been the most widely used photosensitizing reagents in these procedures. However, in their article, *Buckminsterfullerene and Photodynamic Inactivation of Viruses,*[13] chemists Käsermann and Kempf, working in conjunction with the Swiss Red Cross, explore a new interesting alternative, namely the use of C_{60} —buckminsterfullerene. You will remember that scientists found C_{60} in wood charcoal.

A water-soluble C_{60} derivative was tested for antiviral activity. The compound showed a potent and selective activity against HIV-1 in acutely and chronically infected cells. In part, this was attributed to inhibiting the virus' ability to replicate.

No adverse effects to the cells were observed. Furthermore, none of the eighteen tested mice died within the test period when the compound was injected into their peritoneal cavity. These findings should not be unexpected. Scientists have long known that charcoal impregnated swabs should not be used in viral research, as the charcoal tends to shut down viral activity. This research looked into the action of C_{60} in damaging the envelopes that protect a number of deadly viruses including Hepatitis A, B, C, HIV-I & II, and human parvovirus B19. This latest research represents the expanding field of charcoal in medical technology. But, whereas the average person will not be able to use the concentrated C_{60} available to researchers, the benefits are still resident in the lowly charcoal.

Real Benefits

All these facts will fascinate some of us, but mostly we want to hear something tangible. We want something practical that will meet our needs. How have other people used charcoal that helped them in an emergency or with some chronic condition? Exactly what did they do? Because many of those I have interviewed for their charcoal stories travel, as I have, several of the experiences I share are set in developing countries. Those who live in western nations may be tempted to think that faraway stories are not relevant to their world.

North America

While it is true that several of the experiences did happen abroad, I am surprised at how similar they are to the experiences of others here in North America. Beyond that, more and more people are traveling to foreign countries, and are being subjected to a host of strange new 'bugs'. If travelers do fall victim to disease abroad, they quickly discover how limited emergency care can be in developing countries. We all know from the news that diseases, once only known in faraway places, have stormed our homeland defenses, and are sweeping over our once safe refuge. When Europeans came to America, they brought the measles and chicken pox viruses, which decimated some native populations. Once again, diseases that have been unfamiliar to most of us, and in some cases far more deadly than suicide terrorists, are threatening our borders. Cholera is one such menace, as it once again inches its way back to American shores.

Cholera presents itself with diarrhea and vomiting. The body, having recognized some pathogen, is simply doing its best to eliminate it. Unfortunately the diarrhea can often go too far. The body can quickly become dangerously dehydrated and depleted in electrolytes. Without treatment the patient soon dies. But diarrhea and vomiting are classic symptoms of many diseases besides cholera, and the consequences can be just as lethal. Just to name a few, there is typhoid fever, AIDS, stomach flu, microscopic parasites such as *Cryptosporidium parvum* or *Giardia lamblia*, and food poisoning, and you do not have to travel overseas to fall victim to them. But more and more people do travel, and more and more of them are finding that conventional drug therapy is not that helpful.

La Tourista

In their book *Rx: Charcoal,* the Doctors Thrash tell about the experience of one traveler who got sick abroad only to find no help when he returned home:

> "A prominent Columbus, Georgia, executive was traveling to Mexico when he developed a bad case of tourista [the common euphemism for diarrhea amongst tourists. It is generally caused from E. coli bacterium in unsanitary food and water]. He was having diarrhea two to three times an hour on the first day of his disease, with much discomfort in the abdomen. A doctor treated him with the usual antibiotics and Lomotil. A week later, when he returned home, he consulted his own doctor because the condition had settled down to three to five diarrhea stools a day, with continuing discomfort. He got a change of medication but no change of symptoms. After twelve days of the disease and eight different medicines, he called us, desperately seeking suggestions for a natural remedy. We began a routine of one tablespoon of charcoal in a glass of water, followed by a full glass of water every time he had a loose stool. Within two days he was entirely well... and sold on charcoal. Now he never travels without charcoal, and always takes a tablespoon with the first hint of a symptom."[14]

Then there are the more well-informed travelers, like John and Sharon. They headed off on their trip to Mexico with well-wishers warning them about this and that health hazard. But, tucked into their suitcase, as it is no matter where they go, was a bottle of activated charcoal:

> "We traveled with no worries at all, because we knew we were covered. As a precaution, we took a couple tablets every other day. Then we went ahead and thoroughly enjoyed our winter retreat. Sharon even enjoyed a worry-free fresh cut orange from a street vendor."

For those of you planning to travel, especially to developing countries, I would suggest including charcoal in your first aid kit, and not just for yourselves. Remember there are many travelers who know nothing about the benefits of charcoal who may need what you have.

Appendicitis

Obviously one does not have to travel away from home to become seriously ill. Every single day thousands of people are stricken with some sickness that even the very best medical facilities are at a loss as to how to treat. Betty's profession is nursing. Part of the time she nurses at a famous California hospital, and the rest of her time at home:

> "About three years ago my husband, who is a quadriplegic from polio, and also has a paralyzed diaphragm which requires him to sleep in a respirator every night, developed what appeared to be an attack of acute appendicitis. On admission to the hospital the physical exam as well as the lab tests seemed to indicate that the diagnosis was correct and that immediate surgery was imperative. Because of his grave physical condition, he was considered to be a very poor surgical risk. The doctors were not anxious to operate even though they realized that if the appendix ruptured it would be even more hazardous. After several consultations, they decided to operate that night. About eight p.m. the surgeon told us that he would be back in three hours. If my husband was not better by then, they would go to surgery.
>
> "As an RN, I understood the seriousness of the situation. I was devastated at the thought of losing him in surgery or by complications from a ruptured appendix. My previous experience with charcoal as a remedy came to me as a possibility worth trying. After receiving encouragement [by phone] from a friend and a physician, I decided to try a charcoal poultice. I went home and made a large poultice (approximately 14 in. x 24 in.) of charcoal mixed with flaxseed. Unbeknown to the doctors or nurses, I put the poultice on his abdomen for about two hours. When the doctor returned and the symptoms had subsided slightly, it was decided to wait until morning to operate. After the physician left, I put the poultice on my husband again for a while before going home for the night.
>
> "In the morning the doctor saw that my husband was enough improved that they could wait to perform the operation until after the surgeon had completed his scheduled surgeries. Again I applied the poultice. Around noon the surgeon returned, examined him and ordered lab tests. Again the surgery was postponed. I continued to apply the poultice whenever there was an opportunity to do so without getting in anybody's way. Within three days my husband was discharged from the hospital, without surgery!
>
> "After my husband arrived home I continued with the charcoal poultice for ten hours uninterrupted. After three days of this he was not only free

from all symptoms of appendicitis, he was also free from a ten-month problem with liquid diarrhea for which he had tried "everything". Even oral charcoal had not been effective, but the poultice did the job. I must not leave unmentioned the fact that many prayers accompanied the use of this charcoal poultice. It is my belief that when we, in faith, use simple natural remedies, God blesses this effort with supernatural results.

"Please understand that I believe surgery is often the only course to pursue when one has a possible appendicitis, and that it is dangerous and presumptuous to wait too long for treatment. But my personal experience makes me wonder how many times the patient might recover without surgery, if this valuable remedy, charcoal, were better understood and used judiciously.

"It would be a blessing to humanity if some wise benefactor would donate funds for medical research to obtain some statistical findings on the use of charcoal as a remedial agent in medicine and dentistry (as well as veterinary medicine). Meanwhile we can share what we have learned from experience."

It is sharing our experiences with charcoal that opens alternatives to others who are at a loss to know which direction to turn. As we encourage friends or strangers to experiment with this harmless natural medicine they are relieved to know that there is something they can do without endangering themselves or their loved ones.

International

It is interesting to note that the World Health Assembly has urged all member countries of the World Health Organization to promote the use of "traditional, harmless, efficient and scientifically proved remedies" (Resolution 44:34). Charcoal is just such a remedy. It works at home and away from home. It works for local businessmen and for foreign workers.

David is an RN and is keenly aware of the challenges of primitive settings. As the medical coordinator for a non-governmental organization working worldwide, David provides an information sheet on natural remedies for each overseas worker. Most of these travelers are completely unprepared for the challenges of their proposed new homes. Most have been brought up in communities within easy access to emergency and medical facilities. And too, they are largely uninformed about the rational treatment of disease without the use of toxic drugs. David has a varied background. He has worked as a registered nurse in modern American hospitals, as well as establishing health services where no health care existed in a district of fifty villages in remote Papua New Guinea (PNG):

> "While in PNG my project partner John got hepatitis A. He used rest and simple diet to fully recover. It took him about six weeks. I had heard about using charcoal, so when I got the same hepatitis two weeks after John got it, I drank activated charcoal slurry (I don't remember how much). I recovered in two weeks. I also used charcoal at our clinic for simple diarrhea. I would commonly give half a teaspoon of charcoal in water, and then send some home with them to take three times a day. It seemed to be helpful."

With an MPH degree in international public health, David well understands the challenges of working in developing countries. He introduces his information sheet with the obvious:

> "Medicines [pharmaceutical drugs] … require very little effort on the part of the health worker. It is easy to get into the habit of simply relying on medicines. However, medicines have some serious drawbacks:

- All medicines have not only the intended effect but also side effects, some of which can be serious or even life threatening.
- Medicines can make the patient much worse or kill if they are used improperly (the wrong dose, the wrong combination of medicines, or when the patient has a another condition which could be made worse by the medicine—a contraindication).
- By the time medicines arrive at a remote health center the quality may be poor (due to poor manufacturing standards, being outdated, or spoiled by heat or moisture).
- Patients become dependent upon medicines to support their health, but the supply in remote locations is often sporadic. Health workers who can administer the medicines may be nonexistent or only sporadically present.
- Medicines can represent a substantial expense for subsistence villagers.

> "For all of these reasons and more, village health workers should learn to use as many natural remedies as possible, particularly those which rely on locally available resources."

David then zeroes in on his two big guns, hydrotherapy in its many forms (Chapter 17) and charcoal. From first hand experience, David has come to realize just how expertly charcoal works as the sick are nursed back to health. Lastly, he deals with the most common medical problems foreign workers will face: hepatitis, diarrhea (from various causes), boils, abscesses, poisoning, bites and stings. For each, charcoal is a principle treatment strategy.

David's list of common medical problems is certainly not unique to developing countries. These diseases bleed America and other developed nations in billions of dollars of health care costs, and claim tens of thousands of lives annually. The list of medical problems is really much longer. But so too is charcoal's long arm. Before we hear exactly how far reaching charcoal's reputation goes in helping the sick and suffering, let's listen to some who see its chemistry used almost every day. In the next chapter we will begin at the front lines of charcoal's war against poisons, right in our hospital emergency wards. CR

---

[1] *British Medical Journal*, August 26, 1972

[2] *Prevention*, February, p. 136, 1981

[3] Berg MJ, Rose JQ, Wurster DE, Rahman S, Fincham RW, Schottelius DD, Effect of charcoal and sorbitol-charcoal suspension on the elimination of intravenous Phenobarbital, *Therapeutic Drug Monitoring*, 9(1):41-7, 1987

[4] Some have wrongly linked charcoal to carcinogens, no doubt because of their connection to commercial briquettes. Benzopyrenes, which are known carcinogens, are formed when the fat drippings from barbecued meat, poultry, or fish, drip onto the super hot coals and vaporize. These vapors subsequently rise and potentially contaminate the meat. But the formation of these chemicals is dependant on the heat not the source of heat. In actual fact, charcoal adsorbs cancer-producing agents such as methylcholanthrene and benzopyrene, which, when free on the skin, are capable of producing skin cancer.

[5] Hift RJ, Todd G, Meissner PN, Kirsch RE, Administration of oral activated charcoal in variegate porphyria results in a paradoxical clinical and biochemical deterioration, *British Journal of Dermatology*, 149(6):1266-9, December 2003

[6] Pimstone NR, Gandhi SN, Mukerji SK, Therapeutic efficacy of oral charcoal in congenital erythropoietic porphyria, *New England Journal of Medicine*, 316: 390-393, 1987

[7] *Journal of the American Medical Association* 210(10): 1846, December 8, 1969, *Archives of Environmental Health* 1:512, December 1960

[8] Thrash, Agatha, MD, & Calvin, MD, *Rx: Charcoal*, New Lifestyle Books, pp. 60-62, 1988

[9] *Patient Care*, p. 152, October 30, 1977

[10] V. Frolkis, et al., Enterosorption in prolonging old animal life, *Experimental Gerontology*, 19; 217-25, 1984
[11] P. Kuusisto, et al., Effect of activated charcoal on hypercholesterolemia, *The Lancet*, 16: 366-67, August 1986
[12] Kaufman, Richard C, PhD, The Universal Antidote and Detoxifier That Extends Life: Activated Charcoal, *Journal of the MegaHealth Society*, July 1989
[13] Käsermann, Fabian[1], and Kempf, Christoph[1, 2], Buckminsterfullerene and Photodynamic Inactivation of Viruses, *Reviews in Medical Virology*, 8: 143–151, 1998
[1]Department of Chemistry and Biochemistry, University of Bern, Bern, Switzerland
[2]ZLB Cent. Laboratory, Blood Transfusion Service, Swiss Red Cross, Bern, Switzerland
[14] Thrash, *Rx: Charcoal*, New Lifestyle Books, p. 73, 1988

Chapter 8 Poison in the Home

Accidental and intentional poisoning, drug overdoses, and attempted suicide continue to be the most common emergency situations for the application of charcoal in clinical settings. But, there are other emergency applications for charcoal, some more life threatening than others. This chapter begins in hospital emergency wards then moves into the home. We review the recent history surrounding the management of poisoning, and how charcoal has established itself as the premier treatment of choice. We discuss the efficacy of charcoal with respect to different poisons, and conclude the chapter with the story of an attempted suicide in a remote location.

Emergency Rooms (ER)

Pauline is a registered nurse, and has worked in Canada, the U.S. and overseas. She has had many experiences with charcoal in clinical settings, and she chose this unusual one to share. Pauline writes:

> "Charcoal is used routinely in ER's for overdoses and poisoning. I remember this case from when I was working at a hospital in northern California. Sometime after midnight there was a car chase, by the sheriff and his men, after a young man who had a sizable amount of cocaine in his possession. They were finally able to corner him. He tried to make a run for it, but they caught him. However, in order to prevent the police from getting the drugs he had, he swallowed them all. The police brought him directly to our ER. We literally had to pin him down and put a tube down his throat and into his stomach. Then we poured down one and a half bottles of charcoal slurry. He never showed any sign of the cocaine overdose."

As director of the ER, Pauline has witnessed charcoal being used for poisonings, drug overdoses and attempted suicides and can testify to the rapid and powerful action of charcoal. That experience has also come to her aid in treating her own family, as it can for yours. That is, if you have charcoal near at hand.

Dr. Martha, MD, works as an emergency room physician in a Kentucky hospital:

"I use charcoal routinely for poisonings and drug overdoses. I have also ordered a charcoal poultice to be placed over toxic spider bites. We would soak some gauze with charcoal slurry and place that in a disposable hospital chuck [a flat rectangular pad with absorbent material on one side and plastic on the other], and tape it in place with the plastic side out. It worked very nicely on one brown recluse spider bite. I had the staff change it every thirty minutes.

"When another patient of mine called about her brown recluse spider bite, she too was open to using natural remedies. So, I directed her to a friend of mine who lived near her. The friend showed her how to apply a charcoal poultice. She did, and again it healed nicely. I have also used charcoal around our home for bee stings.

"Most people find it too messy, so I will sometimes mix it with KY jelly or hand lotion when applying it externally. If I want to give it to little ones for diarrhea, I may mix it with mashed bananas. I always have it on hand in the home as an antacid, or for any number of common problems."

Ben was, for years, a long distance trucker traveling all over North America.

"I didn't go anywhere without charcoal. If I got an upset stomach, if I had gas, or diarrhea from something, I always had it on hand as a faithful quick remedy. I simply carried a bottle of tablets with me or I would fill some capsules with the powder if I couldn't get the tablets. We keep it right at hand here in the kitchen."

Oh yes, Ben and Martha share the same kitchen.

Poisoning

I have already shared several stories of food poisoning, but one area that is of special concern to parents is accidental poisoning of children. A child is accidentally poisoned every half hour in the United States. Approximately 1.1 million cases, of a child under the age of six years swallowing a toxic substance, were reported to poison control centers in 1998. This number is thought to represent approximately 25% of all such incidents. Things commonly ingested by children less than six years of age include cosmetics, cleaning products, analgesics and cold preparations. Data from 1995 through 1998 revealed that prescription and over-the-counter medications accounted for 52% of the deaths from poisoning during this period. Those substances associated with the greatest risk of death were cocaine, anticonvulsants, antidepressants and iron supplements.[1]

Ipecac

Many older readers may remember one home remedy that was endorsed for many years was ipecac. Ipecac is an emetic, something that causes a person to vomit. The rationale was that if a person had swallowed some poison, then the sooner it was vomited up the less would be absorbed into the body. Of course the very simplest method is to tickle the back of the throat with one's finger, or some other object, to trigger reflex vomiting. Maybe because some people were too squeamish to do that, the product ipecac was promoted as a substitute. But in 1997 the American Academy of Clinical Toxicologists and the European Association of Poison Centers and Clinical Toxicologists reviewed the scientific literature, and issued their new position statement.[2] Among other things they recommended:

- Ipecac should not be given routinely for poisoning management.
- Evidence is lacking to demonstrate that ipecac improves the outcome of poisoned patients and its routine administration in the emergency room should be discontinued.
- There is insufficient clinical evidence to support or exclude ipecac administration soon after poison ingestion.
- Ipecac may delay administration or reduce effectiveness of charcoal, oral antidotes, and whole bowel irrigation.

The American Board of Applied Toxicology and the Canadian Association of Poison Control Centers have since endorsed these recommendations. Only in 2003 did the Academy of Pediatrics stop recommending its use. Suddenly, after decades of aggressive endorsement by the medical professions, we now see these warnings associated with ipecac: "Research has shown that ipecac medication has been improperly administered by parents, and has been abused by people with eating disorders such as bulimia. Abuse of ipecac can lead to heart problems and even death." In light of the above recommendations and considering the fraction of times ipecac is used by parents compared to professionals, it is strange that this new drug alert focuses on parental misuse???

Charcoal

In place of ipecac, the efficacy of activated charcoal used in clinical settings is undisputed. The debate has, however, heated up over its home application.

While there is no question of activated charcoal's superiority over every other known antidote, some suggest it is poorly accepted by young children, making the administering of the recommended dose a little trickier. While this concern has yet to be demonstrated, there are several studies that clearly show the majority of children will accept charcoal. It seems the problems are more with timid parents who are unsure of administering any antidote on their own, and who would prefer to take their children to an emergency department.

Time is of the essence when it comes to poisoning. Charcoal is the treatment of choice for poisoning or drug overdose in children, as well as adults, but it is most effective sooner rather than later. Studies done with volunteers suggest that activated charcoal is more likely to reduce poison absorption if it is given within one hour of ingesting the poison.[3]

Children

In a 2001 study conducted by the Kentucky Regional Poison Center and reported in the medical journal *Pediatrics*[4], Henry Spiller, MS, and George Rodgers, MD, demonstrated the real value of giving activated charcoal **in the home** to children as an antidote for most poisons. The authors noted that activated charcoal is recognized as the treatment of choice when it comes to neutralizing the effects of swallowing multiple poisons.[5,6] [But, its] "use in the home has been limited by concerns that parents would not administer it properly, and that children would refuse to take activated charcoal."

This study, in which activated charcoal was recommended to be given to children in the home, was conducted over eighteen consecutive months and followed 138 cases. Each case was tracked for three days after the emergency call to the Poison Center. Of the 138 cases, 115 (83%) were successfully treated in the home; eight of the mothers preferred to go to emergency; seven could not locate any charcoal in their area; and in six cases the pharmacy was closed for the night; and two could not be followed up.

The time from actually swallowing the poison to taking the activated charcoal was about 38 minutes (±18.3) for home treatment and 73 minutes (±18.1) for emergency department treatment. In 95% of home cases, charcoal was given within less than sixty minutes versus 33% for emergency department management. The average amount of charcoal administered was 12.1 grams. Half (6.9%) as many children who were treated at home vomited after being given charcoal versus those who received treatment in the emergency room (13%). There were no aspirations or other complications.

The authors of the study, Henry Spiller of Kentucky's Regional Poison Center in Louisville and Dr. George Rodgers of the University of Louisville, concluded the obvious: administering activated charcoal in the home is a lot quicker than taking the time to get to the closest emergency room. But, they also said, "Greater efforts need to be put into educating parents about the need to stock activated charcoal in the home in advance of a poisoning." Pharmacists and pediatricians, too, should be made aware that activated charcoal can be used effectively at home.

Other studies done in both America and Europe have confirmed that, "patients at home consistently received activated charcoal in less than one hour after ingestion." In contrast, the majority of patients who were referred to the emergency department experienced a delay of well over an hour. [7,8]

Drugs

As for some specific drugs, researchers have determined that activated charcoal effectively neutralizes ***fluoxetine (Prozac)***, one of the most frequently prescribed antidepressants, and should be administered in cases of its overdose.[9] Researchers also looked at cases of ***acetaminophen (Tylenol)*** poisonings in 981 patients admitted over ten years. In hospital patients admitted within 24 hours, who had swallowed ten grams or more of acetaminophen, those who had been given activated charcoal were significantly less likely (95%) to have high-risk concentrations. Pumping the stomach, in addition to giving activated charcoal, did not further decrease the risk. Researchers concluded: "Activated charcoal appears to reduce the number of patients who achieve toxic acetaminophen concentrations, and thus may reduce the need for treatment and hospital stay."[10] Multiple doses of charcoal, in addition to general supportive measures, are recommended for ***caffeine*** poisoning, with possible further elimination by hemoperfusion (filtering of the blood with charcoal).[11]

Hazards

In one of the very few studies[12] looking at the potential hazards from administering charcoal for poisoning, only two side effects were looked for 1) pulmonary aspiration—accidentally breathing in the charcoal solution or the vomit if the patient upchucks and 2) obstruction of the bowels. Eight hospitals' records were reviewed. Of the 878 patients who received multiple doses of activated charcoal (within a twelve-hour period), only minimal evidence of pulmonary aspiration was identified, and then in only twenty-six patients. Of these, only five (0.57% of total) met all criteria for pulmonary aspiration. None died as a result of pulmonary complications, and none of the patients gave evidence of bowel obstruction. As for aspiration, bowel obstruction or constipation, dehydration, perforation of the esophagus, elevated sodium and magnesium levels said to be associated with charcoal administration, the informed student will discover these supposed hazards are all a result of improperly administering charcoal in hospital settings. Judging from the study in *Pediatrics*, clinicians having trouble administering charcoal, especially to fussy children, would do well to take some lessons from parents who seem to have few if any problems.

Charcoal has also been implicated with the side effects associated with emetics, laxatives, and dehydration. But, failure to provide sufficient water, or the effects of using charcoal in combination with emetics or laxatives, can hardly be blamed on charcoal. As for the complications from aspirating charcoal, any competent clinician will tell you that there is no more danger from charcoal than there is from aspirating the universal thirst quencher—water.

Universal Antidote

When some writers refer to charcoal as a universal antidote, the term 'universal' is not strictly true. Although there are over 4000 substances that are known to adsorb to activated charcoal (see Appendix for a partial list), there are several substances included on charcoal's short list that are poorly adsorbed or not adsorbed at all. They include: Lithium, strong acids and bases, metals and inorganic minerals (such as sodium, iron, lead, iodine, fluorine, and boric acid); alcohols (such as ethanol, methanol, isopropyl alcohol, glycols, and acetone); and hydrocarbons such as petroleum distillates (e.g., cleaning fluid, coal oil, fuel oil, gasoline, kerosene, paint thinner) and plant hydrocarbons (pine oil).

If you are not within reasonable access to an emergency department, it is well to note that while the neutralizing effects of charcoal on the above toxic substances may be minimal, the charcoal will do no harm. In the case of hydrocarbons, it has been shown that very large doses of charcoal will adsorb them in the GI tract. However, in these cases giving charcoal is generally not recommended. Giving charcoal may trigger vomiting which puts the victim at risk of aspirating the vomit which contains the hydrocarbons. Once the lungs are contaminated with hydrocarbons they are quickly damaged. It is supposed, but has never been demonstrated, that the presence of charcoal in the lungs may somehow add to the complications. As a precaution in any case of ingesting a poison, an upright position with the head projected forward poses the least risk of aspirating should the victim vomit.

The presumption that giving charcoal in cases of caustic agents such as domestic bleach, ammonia, methylated spirit, etc., may later complicate the examination of the mouth and throat by a physician is unfounded. Any burns within the mouth or throat caused by caustic agents would be highlighted by the charcoal rather than hidden. As a pathologist, Agatha Thrash, MD, notes that the normal mucosal lining of the mouth and throat is not stained in the slightest with charcoal but easily rinses off. Further, charcoal promotes healing of caustic burns by retarding bacterial growth and inhibiting infection.[13]

While some American pediatric councils have been dragging their feet when it comes to endorsing charcoal, European and Canadian professional bodies have been more progressive. In Canada, the Pediatric Clinical Practice Guidelines for primary care nurses states under GI Tract Decontamination: "Activated charcoal is now recommended as the sole therapy, and should be given for ingestion of any toxic material, except iron, hydrocarbons, alcohols and caustic agents."[14]

Pumping the stomach in a hospital will help to quickly remove large amounts of poisons. However, it does not reach beyond the stomach as charcoal does. As charcoal continues down through the bowels it picks up poisons that were not pumped out. Charcoal and the poisons are then eliminated together when there is a bowel movement. Activated charcoal does not irritate the mucous membranes of the GI system. It may be given orally to someone who is awake and alert. It is generally given as a black liquid drink. Multiple doses of activated charcoal can often be given if someone has swallowed large doses of long-acting, sustained release medication.[15]

Charcoal has the ability to adsorb drugs in the enterohepatic circulation cycle (intestinal-liver circulation loop), significantly clearing them from the blood stream. Normally, the liver filters poisons from the bloodstream. These poisons are then added to the liver bile, and together they pass back into the intestinal tract for elimination. But it is possible for some drugs to be reabsorbed by the intestines and then back into the blood system before they are voided in the stool. Therefore, several days of repeated doses of charcoal is still a good treatment for poisons that may otherwise remain in the blood for a long time.

Charcoal effectively neutralizes alkaloids like strychnine, morphine, nicotine, caffeine, and theobromine.

Every year children die of poisoning from table salt and iron. Because charcoal adsorbs table salt and iron poorly, extra large dosages of charcoal should be used when these substances have been swallowed.[16] If there is any cause for concern, parents can follow up by seeking advice from a physician.

Included in the back of this book is a section **Emergency First Aid** that deals with accidental poisoning.

Aspirin

The most common drug poisoning is from aspirin. In a study reported in the *Journal of Toxicology — Clinical Toxicology*,[17] researchers surveyed 76 poison control centers in North America to compare their recommendations for treating large, acute overdoses of aspirin. Seven toxicologists were also surveyed for informal comparison. Even though there was considerable variation in the recommendations, it was obvious which was their first choice. Seventy-three centers (97%) recommended at least one dose of activated charcoal, as did six of the toxicologists (86%). Only four centers (5%) recommended syrup of ipecac and 38 (51%) recommended gastric lavage. None of toxicologists recommended either ipecac or lavage.

In the case of aspirin poisoning, charcoal should be given right away, or if possible, at least within the first thirty minutes. Powdered charcoal reaches its maximum rate of adsorption in the stomach within one minute. The sooner it is given, the better the chances of successful treatment. After one hour, charcoal given for fast absorbing drugs, like aspirin, is usually only about ten percent effective.

John and Sharon try to be prepared for any emergency. As a quick simple first aid, they always have charcoal near at hand. Because they were prepared for one emergency, they now know it really does work for aspirin poisoning:

> "It happened that a baby got a hold of a number of aspirins and swallowed them. Fortunately it was promptly noticed, and we immediately gave charcoal. For some time after giving the charcoal, we watched the baby carefully, and there were no observable side effects at all."

Mushrooms

How effective is charcoal in cases of mushroom poisoning? More people die from mushroom poisoning than from any other plant. Charcoal easily neutralizes the poison from these mushrooms. For the Amanita (death cap mushroom), conventional treatment with various drugs has had only sporadic success, whereas it is widely known that charcoal neutralizes the poison up to twenty-four hours after ingestion. Charcoal is known to bind the toxic substance alpha-amanitine from water and protein solutions in laboratory experiments.

In one report, seven people, who had eaten more than three mushrooms (three are generally considered a lethal dose), did not seek treatment until sixteen hours later. One of the four women was treated in a separate hospital from the other patients by an exchange transfusion and peritoneal dialysis. She went into hepatic coma for eight days and had to remain in the hospital for three months before she had recovered sufficiently to be sent home. Her husband, who had eaten the same quantity, was treated by hemoperfusion. He left the hospital after six days and was quite well. Four others left after three days.

These individuals were treated with blood-filtering equipment using charcoal, sixteen hours after they had eaten the mushrooms. This would indicate that charcoal can effectively remove the toxins from the blood, even twenty-four hours after eating Amanita.[18] This case suggests that charcoal may also be able to achieve the same benefits when taken orally. You may not have to undergo the risky procedure of hemoperfusion if you have charcoal at hand.

In the Kentucky Poison Center study mentioned above, the authors noted:

> "In appropriate circumstances, rapid, safe administration may allow for prudent treatment of a number of patients who might have previously been treated in the emergency department. In addition to more rapid administration of activated charcoal, home use of activated charcoal produced a considerable health care dollar savings by reducing the number of patients who would have previously required treatment in an emergency department. For example, in our patient group, the 98 patients who had ingested mushrooms would have previously been referred to the emergency for activated charcoal administration and were now able to be treated outside the hospital."[19]

For mushrooms, amanitine toxin content is estimated at more than 5 mg per gram of mushroom. Four ounces of mushroom (a large ingestion for a child) would provide less than 20 mg of amanitine toxin. Two grams of activated charcoal would provide a 100 to 1 ratio. Therefore a five-gram dose (half Tablespoon) would be the minimum dose considered sufficient for successful treatment.

Other Plants

Besides concentrated and synthetic poisons, there are many plants that are also toxic. Here is a list of plants whose toxins are know to adsorb to charcoal: Alder buckthorn berries, autumn crocus, Azalea, black nightshade, bryony, Christmas rose, daphne, deadly nightshade, foxglove, hemlock, holly berry, honeysuckle berries, Jerusalem cherry, laburnum, lily of the valley, mistletoe, monk's hood, oleander, privet berries, rhododendron, savin juniper, spindle tree berries, thorn apple, woody nightshade, and yew.

Experimenting with wild plants to determine their drug effects seems to be a cyclic phenomenon. With the recent hospitalization of several youth in Ohio, Datura inoxia seed poisoning has popped up as another concern. These plants with large fragrant flowers that bloom at dusk are referred to as moonflowers. The seeds are known to have hallucinogenic properties. Some of the poisonous effects also include confusion, agitation, anxiety, seizures, and coma, as well as dry mucous membranes, thirst, flushed face, blurred vision, and fever. Once again, treatment focuses on promptly giving activated charcoal, along with supportive care.[20] Since people will continue to experiment with tasting and eating unknown plants, they would do well to have charcoal on hand.

Together with Food

It has been found that stomach contents can reduce the effective adsorption of charcoal up to 50%. Therefore, when a poison has been ingested, the Doctors Thrash take the following strategy:

> "To be on the safe side, use, as the dosage of charcoal, approximately eight to ten times the estimated weight of the poison. Finely powdered charcoal can get to the surface of toxins better than coarsely powdered charcoal and should be used for the best results."

But, there is a good chance we might not know the type of drug or poison or the quantity taken. So, other doctors recommend a fixed amount of 50 to 100 grams. In their classic study, Hayden, PhD, and Comstock, MD, were emphatic, "In reality, the dose of activated charcoal that should be administered to adsorb drugs from the GI tract is in the range of 100 to 120 gm."[21] Because they understood that there is no worry of overdosing, these doctors, who had administered charcoal in hundreds of cases, were not afraid of larger doses. One tablespoon of charcoal equals about ten grams. Fourteen capsules equal about a tablespoon of powder.

Dosages

So, what is the optimum dose of activated charcoal to be given? As you can see, there is no clear consensus among those promoting charcoal in cases of poisoning, except that you can't give too much. There are no definite dosages, but in his book *Activated Charcoal in Medical Applications*, Cooney noted these three recommended formulas based on:

- Age—25gms to 50 gms for children
- Body weight—1gm per kg body weight
- Amount poison taken—10 gm of activated charcoal per 1 gm of poison.[22]

Emergency Kit

Although we do our best to keep dangerous substances away from our children, they still sometimes get hold of 'sweets', like aspirin/antipyretics, sleeping tablets or other drugs or 'lemonades', like silver polish, washing-up liquid, bleach, and insect repellent. So, an informed attitude to preparedness will mean including activated charcoal in your emergency kit. In addition to the powdered charcoal, you should have a bottle of capsules or tablets.

Commercial bottles of USP activated charcoal are also available and are considered adequate. Usually for less than $5 (U.S.), you can purchase plastic containers with 15 grams of dry activated charcoal. They come as either a slurry of charcoal pre-mixed in water, or as a container into which water or soda pop is added. If they do not already have it, ask your local pharmacist to stock charcoal.

How to Give Charcoal

A convenient way to have a charcoal mixture available when you need it, is to put thirty grams (three Tablespoons) of charcoal into a glass jar with a screw on lid. All you have to do when it is needed, is to add four ounces of water, shake the jar a few times, remove the lid, and drink the contents. Follow this with one glass of water. Of course, if you are fortunate enough to be near a pharmacy that is well stocked, you can purchase the commercial product that Pauline used in the ER at her hospital. In an emergency, parents will have greater success if the children have already seen adults using charcoal. Nevertheless, if parents will show confidence and firmness there will be little resistance.

If the charcoal is vomited, give another dose immediately. Follow the dose of charcoal with a full glass of water if possible. Charcoal taken the third time will usually stay down.

For Beginners

For beginners, I would suggest the use of capsules, as they are more likely to be activated, more adsorptive, and easier to swallow. But some brands of tablets are slightly sweetened and/or color coated. I personally think that is a great idea. Whatever is lost by adding sugar is more than made up for if the child takes it willingly because it is sweet. Remember, by itself charcoal is gritty, tasteless and black. Not the most appealing to a child. So learning some camouflage techniques can be really helpful. I also prefer capsules because they are convenient and non-messy. Tablets can take a little longer to dissolve. Remember too, charcoal tablets are a little more than half as adsorptive as powder, so you will need about twice as much. Powder is the best way to quickly ingest a large dose, but can be messy. However, the fact that it temporarily blackens the teeth, gums and tongue may add some humor to a tense situation.

Once again, as a general rule, a single large dose of activated charcoal should be taken as soon as possible after swallowing a poison. Remember, swallowing large amounts of charcoal is harmless, and taking too little is ineffective. In an emergency few people will be able to precisely determine the amount of the poison. The actual effectiveness of the charcoal will vary, so take more than you think you need. Charcoal should be taken within thirty minutes of swallowing the poison. The longer the delay, the less effective charcoal will be, losing up to 60% of its usefulness.

So, whether you are waiting for an EMT, or you are driving your patient to an ER, be sure to give the victim every benefit by quickly preparing and giving a substantial dose of charcoal.

Further Action

Depending on the poison and the amount taken, as well as one's access to a phone or transportation, a call or emergency visit to the hospital may be advisable. Otherwise, if the patient is conscious, induce vomiting by pressing one's finger or a spoon handle against the back of the tongue (Do not do this if the patient has swallowed a corrosive or petroleum-based substance). If the patient is unconscious or appears to be losing consciousness, turn him on his side, make sure that his air passage is unrestricted and prevent him from vomiting. If possible, take the remains of the poison and/or its packaging with you to the hospital to show to the doctor. Also, because a small percentage of patients do vomit after being given charcoal, having a waste container at hand may be practical.

Poison Control Centers

In the United States the universal telephone number for the Poison Control Center is **(800) 222-1222**. Calls are routed to the local poison control center. In Canada the number can be found within the front cover of your phone directory. Of course, if your area has it, you can always call **911**.

Unconscious

Giving charcoal orally is not recommended if the patient is unconscious. That is, of course, if you have the option of a nearby ER or competent emergency medical technician (EMT). If not, what are your options? I refer you back to my very first experience in Guatemala when Dr. Graves applied a charcoal poultice to a semi-conscious girl. Applying a generous amount of charcoal paste to the abdomen may not look very tidy, but if your options are very limited it will do no harm and may bring just as prompt a recovery. Immersing an unconscious victim in a charcoal bath may be another alternative.

There is also the case mentioned earlier, of the unconscious young child in Nepal who was given a charcoal slurry by a pediatrician and nurse. The case that follows tells of an attempted suicide, who also completely recovered after being given charcoal orally by relatives. It should be noted that there were no emergency facilities available for either the doctor or the laymen.

General Detoxifier

While most people usually think of poisoning as a single event, we need to be reminded that the body is continually dealing with waste products from digestion and metabolism. The organs employed in filtering and purifying the blood can always use some extra help.

As a general detoxifier, charcoal is without equal. It purifies the six to eight liters of digestive fluids that are secreted daily. This in turn helps to remove foreign substances from the blood. Charcoal adsorbs the intoxicant substance and its metabolites that are excreted into the small intestine by way of the bile duct, thus preventing their reabsorption. As we have noted, charcoal adsorbs drugs that diffuse back into the stomach and intestines. By neutralizing all these toxins, charcoal decreases the workload of the liver as well as the kidneys. While charcoal lightens the load on the body, it could also lighten our national budget.

Financial Savings

Hayden and Comstock made these rough calculations back in 1975: "Assuming the figure of ten intensive care admissions per day (for acute poisoning) per 2 million population and assuming Houston to be an average city, a very conservative estimate is that there are 1,825 admissions per million population per year. With the population of the United States being approximately 200 million, an estimated 365,000 admissions to intensive care units each year in the United States are due to drug overdose." They felt this was a very conservative figure. Based on their clinical experience, they estimated the average intensive care stay of three to four days for severe drug overdose could be reduced at least by one day by giving activated charcoal. Reducing the intensive care days by 365,000 annually at a cost estimated to be about $200/day, netted a saving of $73 million per year. This does not include the reduction in suffering and death.[23] One can only guess what multiple millions more could be saved today with the increase in population, skyrocketing emergency care costs, and adjustments for inflation. Charcoal may not save you millions of dollars personally, but after you are saved one trip to the ER, you will be counting more than your lucky stars.

At the time of their research (1975) the authors noted, "Activated charcoal is a safe, effective, and inexpensive gastrointestinal adsorbent that is recommended for use by virtually every recent textbook or handbook on the treatment of acute drug intoxications. It is paradoxical, then, that this antidote is seldom employed in the clinical practice of emergency medicine. In fact, activated charcoal as a recognized drug is not available in the United States, and commercial preparations of the product usually specify that it is not to be used in human subjects. Hopefully, this proscription against activated charcoal may be removed in the future."[24]

Here in the 21st century we can be very grateful for those visionaries who pushed against the prejudice of their day. Today we see activated charcoal made available in many pharmacies and health food stores. The tide has definitely been turned, and other progressive-minded researchers, doctors, educators, and nurses are using and aggressively championing the powers of charcoal.

Nausea and Vomiting

The Doctors Thrash have repeatedly treated nausea and vomiting with charcoal. They write: "We have never seen a case of acute nausea and vomiting in which treatment was begun early that continued past three doses of charcoal kept down. We believe it to be the most effective treatment available for nausea and vomiting and should always be used as the primary treatment of choice. Often, very uncomfortable patients will feel well in seconds after swallowing the charcoal slurry made from charcoal powder stirred in water."[25]

In the case of nausea and vomiting, after each vomiting episode, drink a mixture of one to two large spoonfuls of powder in a little water. You may follow the charcoal water with a glass of plain water, if tolerated. If the charcoal is vomited up, drink another glass of charcoal water immediately.

For diarrhea, after each loose or watery stool, drink a mixture of one to two large spoonfuls of charcoal powder in a glass of water. Follow each glass of charcoal water with one or two glasses of plain water.

Abortions

In the 1930 edition of *Surgery, Gynecology, and Obstetrics*, Nahamancher, MD, reported the effectiveness of charcoal in lowering the fevers associated with septic endometritis in women who had undergone abortion. In almost all cases the temperature fell to normal within twenty-four hours when one or two "charcoal pencils" were inserted into the uterus through the cervix. In those cases not treated with charcoal, a low-grade fever persisted for several days. Dr. Nahamancher naturally assumed that, for those women treated with charcoal, it had adsorbed the offending toxins and bacteria in the uterine cavity. Once he started using the intrauterine charcoal, he never lost another case of septic abortion. He also reported that in cases of infected abortions, when treated with charcoal, there was an almost immediate end of the foul-smelling discharge.

Dr. Nahamancher further reported that the fever sometimes associated with childbirth was likewise very favorably influenced by intrauterine charcoal therapy. When a foul smelling discharge and low-grade fever set in, the patients were treated for five to six days with conventional practice. If no improvement occurred, charcoal was introduced into the uterus on the seventh day under the strictest aseptic precautions. In all cases a reduction in temperature occurred the day following and was normal by the second day. The odor disappeared at once. In over 90% of cases it was unnecessary to insert charcoal pencils more than once.

His method of making the charcoal pencils is interesting. He heated water, charcoal and starch together until thickened. After slight cooling, he poured strips of the tacky material onto a greased pan. When completely congealed, he placed them in a barely warm oven overnight to dry. He could then pick them up if he handled them gently. [26]

Pesticide Poisoning

Before going to the next story, the reader needs to be reminded: Do not give charcoal orally to an unconscious patient, unless… Unless you happen to be in the middle of nowhere, with no other remedies, and with no emergency help available.

Ruth RN was traveling and lecturing in northern India in the province of Mizuram presenting public health classes. In her series, she also makes mention of simple home remedies, and naturally talks about the benefits of charcoal. After Ruth had returned to the United States, her translator conducted his own programs. An elderly man who had attended the meetings later returned to his village.

As Ruth related the story, this man's daughter came home one day to find her husband unconscious in bed, with a container of pesticide beside him. The wife realized that he must have taken a large amount of the poison intending to end his life. Frantically she ran to her nearby father's house, begging for help. Instantly charcoal came to the old man's mind. He quickly had everyone go to the neighboring houses and gather up charcoal from the kitchen fires. He then pounded it as best he could. He mixed it with a cup of water, and they proceeded to force it down the husband's throat even though he was unconscious. It took a great deal of effort and time, but after working on him continuously for several hours, he began to show some resistance. They continued forcing it down him little by little until he finally revived. It was five to six hours before he regained full consciousness. As the days passed there appeared to be no adverse effects from the large dose of poison he had taken."

A remarkable story! As Ruth told me, this kind of experience very seldom happens in western countries. Because of the availability of emergency clinics, and because of the general ignorance of first aid measures and the efficacy of charcoal, Americans are predisposed to take their chances and wait until the ambulance arrives. But for the differences there are similarities too.

As in India, so in America, there is a great need for education. Not only do laymen need to be informed, educators need to learn as well. It is easy to listen to stories. It is easy to forget the stories. It is easy to put off buying charcoal for your medicine chest. It is easy to leave our health to others. But in the end it is our loved ones or ourselves who may suffer.

In Review

In case of poisoning:

- Immediately drink 4 to 10 large spoonfuls of charcoal mixed with a small amount of water.
- Fill the glass with water and drink the contents with the remaining charcoal sediment.
- If a person has eaten within the past two hours, more charcoal will be required.
- Repeat the charcoal dosage in ten minutes, and any time symptoms worsen.
- On the safe side, the patient should be taken to the hospital as quickly as possible for monitoring.
- As a general rule, treat children with one half of the adult dose.

When stored dry in a dry, tightly covered container, charcoal powder, tablets and capsules will keep indefinitely. Always keep a container of charcoal powder in your first aid cabinet so that it will be readily available for the unexpected.

Charcoal is not a panacea, but it is one in an arsenal of simple rational strategies. With care and diligence, it has often brought relief to some sufferer. The same could be true for you. CR

---

[1] Kirchner Jeffrey T, Update on the Management of Childhood Poisoning, *American Family Physician*, June 1, 2000

[2] Krenzelok EP, McGuigan M, Lheur P. Position statement: ipecac syrup. American Academy of Clinical Toxicology; European Association of Poison Centres and Clinical Toxicologists, *Toxicology*, 35:699-709, 1997

[3] Tenenbein M, Green R, Grierson R, Sitar DS, How long after drug ingestion is activated charcoal still effective? *Journal of Toxicology & Clinical Toxicology*, 37:610, 1999

[4] Spiller, Henry A, MS, DABAT, Rodgers, George C Jr, MD, PhD, Evaluation of Administration of Activated Charcoal in the Home, *Pediatrics*, Vol. 108 No. 6 December, 2001

[5] Bond GR. The poisoned child. Evolving concepts in care, *Emergency Medical Clinician, N. Am.*, 13:343-345, 1995

[6] Chyka PA, Seger D. Position statement: single-dose activated charcoal, American Academy of Clinical Toxicology; European Association of Poison Centres and Clinical Toxicologists, *Toxicology Clinical Toxicol*ogy, 35:721-736, 1997

[7] Lamminpaa A, Vilska J, Hoppu K, Medical charcoal for a child's poisoning at home: availability and success of administration, *Human Experimental Toxicology*, 12:29-32, 1993
[8] Wax PM, Cobough DJ, Prehospital gastrointestinal decontamination of toxic ingestions: a missed opportunity. *Journal of American Medicine*, 16: 114-116, 1998
[9] Atta-Politou, J, et al, An in vitro evaluation of fluoxetine adsorption by activated charcoal and desorption upon addition of polyethylene glycol-electrolyte lavage solution, *Toxicology Clinical Toxicology*, (36(1-2):117-124, 1998
[10] Buckley NA, Whyte IM, O'Connell DL, Dawson AH, Activated Charcoal Reduces the Need for N-Acetylcysteine Treatment After Acetaminophen (Paracetamol) Overdose, *Toxicology Clinical Toxicology*, 37(6):753-757 1999
[11] Nagesh RV, Murphy KA, Caffeine poisoning treated by hemoperfusion. *American Journal of Kidney Diseases* 4: 316-318 1988
[12] Sadovsky Richard, Complications of multiple doses of activated charcoal, *American Family Physician*, Oct 1, 2003
[13] Thrash and Thrash, *Rx: Charcoal*, p.58, 1988
[14] Pediatric Clinical Practice Guidelines for Primary Care Nurses, Health Canada, 20 pp.13-14 July 2001
[15] Huston, Craig C, MD, (Staff Physician, Department of Emergency Medicine, Cook County Hospital, Rush University Medical College), co-author Bilkovski, Robert N, MD, (Consulting Staff, Department of Emergency Medicine, Christ Hospital and Medical Center), Activated Charcoal, *eMedicine.com, Inc.* Nov. 30, 2003
[16] Thrash, Agatha, MD, & Calvin, MD, *Home Remedies*, New Lifestyle Books, p. 144, 1981
[17] Juurlink, DN, McGuigan, MA, Gastrointestinal decontamination for enteric-coated aspirin overdose: what to do depends on who you ask, *Toxicology Clinical Toxicology*, 38(5):465-70, 2000
[18] Wauters, JP and Rossel, C, et al, Amania phalloides Poisoning Treated by Early Charcoal Hemoperfusion, *British Medical Journal p. 1465;* (comment by Dr HR Brunner), Nov. 25, 1978
[19] Spillar, Henry A, MS, DABAT, Rodgers, George C Jr, MD, PhD, Evaluation of Administration of Activated Charcoal in the Home, *Pediatrics*, Vol. 108 No. 6, December 2001
[20] Carrie Morantz, Brian Torrey, CDC information on moonflower intoxication. (Clinical Briefs), *American Family Physician*, Nov 1, 2003
[21] Hayden, James W, PhD, Comstock, Eric G, MD, Use of Activated Charcoal in Acute Poisoning, *Clinical Toxicology,* 8(5) p. 523, 1975
[22] Cooney, David O, ed. *Activated Charcoal in Medical Applications*, New York, NY: Marcel Decker; 163-184, 1995
[23] Hayden, James W, PhD, Comstock, Eric G, MD, Use of Activated Charcoal in Acute Poisoning, *Clinical Toxicology* 8(5) p. 530, 1975
[24] *Ibid.*, p. 515
[25] *Rx: Charcoal*, p. 40-41
[26] Surgery, Gynecology, and Obstetrics, 96:873-878, 1930; See also *Home Remedies*, Thrash & Thrash, MD, pp. 151-152, 1981

Chapter 9 At Home on the Job

Charcoal works for science and technology. It works for the food industry and the medical industry. It works on land, under the oceans, and up in outer space. It has worked for thousands of years, but I suppose the place it is used most faithfully, most affectionately and most gratefully is right in our homes.

On the Job

As we have seen, charcoal is not some obscure treatment for the more primitive places of earth. So, when I listen to people's first-time experiences with charcoal, especially here in North America, I am somewhat bemused, but not totally.

Risë is first a homemaker, but she also writes a health column, and has appeared on TV numerous times with her health nuggets. She says, "My first experience with charcoal was back at Pacific Union College [California] long before I knew about natural remedies. I had severe abdominal pains that left me in bed for three days. I finally got up enough energy to go to the hospital. The doctors didn't know what was wrong with me, but they thought that maybe I was suffering from appendicitis. The pain did not go away. Another student I had met some time before came to visit me, and brought a Tupperware cup filled with this black stuff. I looked at it, and then at her. She said, 'Don't even bother asking. Just drink it.' By then I was willing to try anything. So I drank the whole thing. After having suffered for days, I was feeling better in about half an hour! I thought 'What is that stuff, hocus pocus?'

"I have used it since on the run-of-the-mill little emergencies like bee stings, or upset stomachs. I have found what works well for the children is a fizzy drink with charcoal mixed in, and a straw to drink it with. Anything fizzy is good, but fruit juices don't work for us. What seems to work best? James prefers capsules, but I feel the powder works better, so the kids and I use that."

When I set out to gather these stories, I already had a variety of friends and acquaintances to draw from. Sometimes they referred me to others. I also found some stories about charcoal on the Internet, and people who were willing to have them shared. Whenever I had a lead, I would try to follow it up. If I suspected people might be inclined to use natural remedies, I would simply ask if they ever used charcoal. Of those that did, most used it only for stomach upset or gas. But again and again, I would chance upon some folks who really do use charcoal for a variety of ailments.

Erna and Kelly are veteran charcoal users. Erna runs a busy household with two teenagers. She sent on some of their history:

> "My family has been using charcoal in various ways for many years. At first, we mostly used it for eliminating intestinal gas, but as the years have gone by we've discovered other useful benefits as well. When Sarah was in the hospital with a broken femur (thighbone), she had to be in traction for two weeks. The pain medication she was on (codeine) reacted on her digestive system and shut it down completely, causing her a great deal of discomfort. The nurses and doctors offered no remedy, and she suffered for two or three days. Finally, I asked Kelly to bring in the charcoal and I gave her a dose of it (one heaping teaspoon in a glass of water). She drank it, and within an hour the discomfort was gone. The nurse was quite amazed."

Whether as a folk medicine or as a clinical treatment, charcoal enjoys a reputation that is worldwide. Actually, it seems that it is the general North American public that is uninformed when it comes to the varied uses and benefits of charcoal. And yet, we are, in fact, surrounded by it. A day does not pass that most of us are not benefited in several ways by charcoal.

One large manufacturer of charcoal in China is tapping into the universality of charcoal as it aggressively markets a number of charcoal products. They recommend charcoal as a decontaminant in treating domestic water to eliminate chlorine, carbon dioxide, and other saturated gasses. Removing these gases naturally increases the alkalinity of the water, which makes washing things easier, and makes for smoother skin. If added to water in a flower vase, they claim it will prevent the stagnant odor that often develops, improve the water absorption, and prolong the blooming of flowers. Turning some into potting soil is also beneficial. While charcoal has long been used in aquarium aerators to purify air, they also recommend sprinkling some in the water to help kill viruses and bacterial growth. It can be used anywhere in the house where there is a problem with odor, mildew, or mold. The company also manufactures cushions that they claim are a benefit to long distance drivers and blankets layered with charcoal cloth that improve sleep at night.

Attitude

Some will find the above list of charcoal's varied uses a little hard to accept. As with Risë and myself, most people typically have an attitude reaction. Consequently, many veteran charcoal users have found that, when trying to introduce charcoal to those who are unacquainted with its benefits, sharing one's own experiences often works as a starting point. If there is some health problem that might respond to charcoal, then a few scientific explanations, some genuine sympathy, and a simple demonstration of how to take it, will go a long way to winning over a doubter. If you are new to charcoal may I suggest first trying tablets or capsules. Not because they work better, but because stirring a tablespoon of black powder into water, and then offering the smoking brew to a beginner will almost certainly trigger an "attitude problem".

Attitude is important, and when promoting charcoal, because of our natural prejudices, a positive picture is especially helpful. So, when trying to help others explore a remedy outside their familiar medicine box[1], sometimes "less is more". "Here, try this". Whether it's gas, heartburn, acid indigestion, or a "sore tummy", it is not uncommon to hear the afflicted one say, "Hey it works, what was it you gave me?" "Oh, just charcoal." "Did you say charcoal?"

In the Store

On most commercial containers you will find something to this effect: 'This product is derived from a special high-quality grade charcoal.' Or, 'Contains USP (United States Pharmacopoeia) activated charcoal[2], vegetable sources and binders', or something to that effect. Typically the distributor will make no claims. The exception may be, 'Indications: symptomatic relief of gas or indigestion'. Dosage will read: "One to two capsules or two to four tablets as needed." Since there has been no recorded instance of overdosing on charcoal in all the literature during the past century, (or the past four millennium for that matter), there are no warnings to that effect.

Some labels explain that the stool will darken temporarily and, because of charcoal's tremendous adsorptive qualities, it may interfere with the effectiveness of some medications if they are taken too close together with charcoal—within two hours of each other. Studies have shown that charcoal's adsorptive powers diminish quickly after two hours from ingestion. After taking medication, most medical literature suggests a two-hour wait is sufficient before again taking charcoal. The label may also suggest that the charcoal be taken after or between meals. Because of its binding nature, some distributors remind users who are susceptible to constipation, to drink sufficient water. One thing you will not find on the bottle or elsewhere, as you always will with drugs, are warnings of adverse effects. Again, four thousand years of history has yet to conclusively demonstrate any direct adverse reaction. Certain monkey colonies have eaten charcoal on a regular basis for at least that length of time, and they too have never manifested any negative reactions.

Can someone overdose on charcoal? Let me ask, can someone overdose on water? In most cases it is called drowning. There are the odd individuals who drink so much water that they seriously dilute their electrolytes, and bring upon themselves as many problems as they are trying to flush away. Common sense will guide people who are health conscious as to how much water they should drink to stay well hydrated. Common sense will also guide them as to how much and how often to take charcoal. The danger of overdosing on charcoal is far more remote than overdosing on water. Unlike water, charcoal by its very nature, does not lend itself to overindulgence, either because of taste, texture, or biochemical stimulation. There are no recorded cases of overdosing on charcoal anywhere in the scientific literature!

In the Home

As you read through the following pages, you will hear some strange and amazing stories about charcoal's ability to help restore health. However, by and large, it is the day-to-day digestive upsets and "up chucks" that make up charcoal's reputation.

Dianne's experience is that of most who have used this age-old remedy. She writes from British Columbia, Canada:

> "As for charcoal, we use it all the time. As you know I am Dean of Girls here at Fountainview Academy, and I wouldn't be without charcoal. For upset stomachs, sore throats, earaches, and for most anything else that comes along, I hand out the charcoal or make charcoal poultices. They always work. As far as anything spectacular that I could tell you about, I can't think of anything in particular. Well, just last Wednesday I took some girls shopping. On the way home we stopped for supper, and one of the girls got stung by a hornet. We really had nothing handy to take care of it, so she suffered until we got home. I then mixed up a warm charcoal poultice, and put it on her thumb. She kept it on overnight. The next morning the swelling had all gone, and she was feeling no more pain."

Then again, there are the unusual cases that come along, sandwiched in between the more typical emergencies.

Knowing just enough of the challenges of their next home, mixed with some apprehensions, Jeff and Bobbi agreed they better be prepared for any unforeseen emergency. When they decided to leave their familiar surroundings in America and make their new home in the unfamiliar jungles of Papua New Guinea for several years, they took along a fifty-pound sack of activated charcoal. It wasn't long after they arrived that it was called into duty. Bobbi writes:

> "We used charcoal shortly after arriving in the village. Pasiyata [one of the local women] had gone to gather sago—a common food staple of the area—and had gotten a sago thorn in her upper thigh. The thorns from the sago tree are very long and very sharp. Even after we removed the dead skin over the wound, there still was no sign of the thorn. It was very deeply imbedded, and was infecting the surrounding area of her leg. Every day she would come to our house in the morning and evening. We did hot and cold treatments, and then we wrapped her thigh in a charcoal pack. After

> about one week the charcoal drew out the thorn—all two inches of it. From that day on we were promoted as people with powerful medicine."

Powerful but very affordable and readily accessible because, as in most developing countries, charcoal can be easily made, and it is a common cooking fuel.

Later Jeff got an infection in his leg too and they turned to charcoal. In his letter to me Jeff wrote:

> "When my leg got infected, we used charcoal poultices to draw out the infection. In the end we still had to go to the Highlands and get antibiotics to get my leg to clear up. It was the charcoal though, that had kept the leg from getting too bad until we could get medical treatment."

While working on the manuscript for this book, I took a break and wrote an article on charcoal for a Canadian magazine. It included some of the stories you will read here, and it gave some simple instructions on how to take charcoal. Sometime after my friend Jared happened to see the article, he told me that, about that time, a violent stomach flu had run through the area. He had had his family take charcoal and none fell victim. However, several of his church family were not so fortunate.

Previous to this, while getting a haircut one day, Jared became acquainted with his hairdresser, and their discussion revolved around natural health. She had some misgivings about the healthcare system. Later, she developed a severe sinus infection, nagging cough and congestion. She went to several doctors who prescribed different things. They would help for a while, but the symptoms would soon return. When the hairdresser told Jared of her health problems, because the article on charcoal was still fresh in his mind, he immediately went and got the article and some charcoal and brought it to her.

She had planned on going to the outpatient clinic because her doctor was on duty, but decided she didn't want to wait a couple hours to see him. After reading the article, she decided to give the charcoal a try. She took several capsules, and went to bed. When she woke up the next morning feeling better, she knew it was the charcoal that had made the difference. She gave the article to her husband and now they are both convinced. Like Jared, you too can have the joy of recommending charcoal, knowing it may help some sufferer and, if not, it will do no harm.

Charcoal is just as powerful, whether it's used in some foreign jungle, a Canadian hamlet, or a large urban center. Humphrey operated a large construction company for many years in a large city. He and his "nurse" Edith have used charcoal for many years. Edith told me over the phone, "If we eat out and something doesn't sit well, or if we feel a cold coming on, we will take a couple tablespoons of activated charcoal and things settle right down. Humphrey had a troublesome pimple on his back, so

I made a charcoal poultice with flax seed, and put that on overnight. Worked like a charm! Sometimes I mix the charcoal with bread to help it stay together. For sore throats we gargle with charcoal water or we'll suck on a charcoal tablet." Simple—simple to give, simple to take.

Many older folks have vivid memories of some simple home remedy and seldom have any reservations about using them. Ed's mother is eighty-eight years old. She took a tumble and cut the middle finger on her left hand. While I was visiting at their home, Ed told me what they did:

> "It was a nasty gash that got badly inflamed. It swelled to twice the normal size. The doctor thought the overextended tendons contributed to the swelling. We tried hot and cold soaks with Epsom salts for several days, but she did not like the hot part. The infection got worse and became an angry red. So we went and purchased some charcoal capsules at the store. We broke open several capsules, mixed the powder with some water, and placed the soaked charcoal in a cloth. Then we wrapped the cloth around the finger, and left it on overnight. By the next day it was noticeably improved, and it continued to heal."

It matters not whether you are old or young, charcoal promotes healing. When Dale was about thirteen years old he developed severe ingrown toenails. The toes had to be operated on. The nails and root were removed and the nail beds stitched. Before he was to return to the doctor to have the stitches removed, the toes turned pink, and then became inflamed. The toes became shiny from the swelling. His doctor prescribed antibiotics, but his parents decided against them. Instead, they tried a charcoal poultice. In the early evening, just before going to bed, they took dry white bread and put it into hot water to soften it. To this they added one tablespoon of charcoal powder. This was mixed together into a paste while still hot. Then it was left to cool. Sheila, Dale's mother told me, "I placed that directly on the stitches, wrapped plastic around the toes, and put a wool sock over that.

Dale woke in the night with the toe throbbing. So, I removed everything, bathed the foot and made another poultice. When I removed the poultice in the morning, not only had it drawn out pus, it had drawn out the stitches too. I washed the foot again, and placed on another poultice. The foot still throbbed. On the morning of the fourth day it began to look good, even though the toes were pretty black. When I took Dale back to the doctor and he removed the sock he was amazed to see so much improvement. He said, "Dale you're lucky you have a wise mother. Otherwise I may have had to lance the toe to remove the poisons."

Wise mothers are wise even when they are not around. Erin did some volunteer work in Mali, West Africa, last year. Her mother Cathy is a nurse, and encourages the use of simple remedies. So Erin has been raised knowing the virtues of charcoal. I asked her if she had anything she would like to share after having spent a year making her home away from home:

> "We certainly kept charcoal around, but I don't remember any great stories about it in particular. Well, there was one time that it came in handy to me. I was working with a friend, and since both our parents have all these great stories about charcoal, we were pretty familiar with its interesting properties. Paul was being a little sarcastic telling me how people always use charcoal as a cure-all. He felt some people will stuff charcoal into you, when you're feeling sick to your stomach, as though it will certainly keep everything in your stomach. I've probably thought that very thing at one time or other, but what I said out loud was that my mom always said that it was good to take charcoal anyway, because if you threw up, it would neutralize some of the stomach acids and do less damage to your esophagus.
>
> "We happened to be trekking through the African bush one Sabbath afternoon, hoping to see one of the rumored leopards (we never did). I was sucking down water as though my life depended on it. It wasn't really working though, because what it felt like was that, instead of being absorbed into the rest of my body, all this water was sticking in a big, sloshy glump in my stomach. Finally we turned around and went home.
>
> "Later that evening, I stopped by Paul's hut to use some of his charcoal—I didn't care what he said, I wanted my charcoal. Sure enough, the big glump in my stomach fought its way back to the surface, if you know what I mean. And also sure enough, I was glad to have taken the charcoal. Instead of that awful post-vomit burning, I went straight back to bed and slept peacefully the rest of the night. I still don't know why my stomach didn't want any water that afternoon, though."

Maybe Erin should ask her mom. It seems mothers know best.

But, I have my own questions. If Paul was so skeptical why did he happen to have charcoal with him? Did his mother pack his backpack? And, if Erin was sold on charcoal, why did she not have any of her own? Did her mother forget to send her off with some?

Urinary Tract Infection (UTI)

Martha sent this story on to a friend who, knowing about my project, sent it on to me. It is a heart-warming story:

> "I'd love to share a bit of information with you regarding my daughter's problem of recurrent urinary tract infections (UTI). She is seven and

one half years old now. As I recall, this started shortly after she was toilet trained, about two and a half years of age. She developed severe burning upon urination, frequency, urgency, and bedwetting. The doctor put her on antibiotics, and the problem would clear only to recur again and again. Her fever would go up, and, with subsequent episodes, began to develop pain in her abdomen and sides. The infection would quiet down a bit, but never totally went away. The doctors changed antibiotics, but she never could get totally clear of the bacteria in the urine.

"She finally had a surgery in which the doctor scraped the lining of the bladder, dilated the urethra, and checked the ureters and kidney areas for signs of the infection there. She had a great deal of scar tissue built up in the urethra and bladder from all the infections over the years. This summer the doctor told me that she would probably need to have further surgery to try to correct these recurrent infections. He felt that perhaps scraping the bladder again might reduce some of the infections. He said that probably she would need to be on antibiotics continually as there was no other answer to controlling the infections.

"When you shared with me the solution of using charcoal in water, and letting her drink this, I went home and mixed the solution as you directed. She was symptomatic for UTI at the time, and her abdomen was hurting badly. Within one and a half hours she said the pain was gone, and by the following day, the urgency and frequency was gone also. We use the clarified charcoal/water mixture daily now, and to date she has not had a single recurrence of the infection!

"Such a simple solution to a complex problem! I only wish we had learned of this years ago. All the pain and the expense would have been nonexistent! Her father is a respiratory therapist and I am an RN, and so many times now in the hospital we see patients that we are quite certain would benefit from the use of charcoal. Unfortunately though, the advances of medicine lead us far away from the simple cures, no matter how effective they may be.

"I hope you can share our experience with others so that the word will get out quickly about the 'miracle drug'—charcoal. Thanks again, Martha"

Crohn's Disease

Barbara has been something of a TV celebrity here in America. Several years ago she found herself in front of the camera because of her Crohn's disease. She had learned how to combat her illness using a holistic approach, and as a consequence developed her cooking skills to make what otherwise would have been a very bland diet into a gourmet's delight. A worldwide satellite program featuring different health-related subjects invited Barbara to demonstrate some of her creations. Little did the program producer realize how contagious her vibrant enthusiasm would be. Not only did hundreds of people identify with her disease, but thousands more fell in love with her and her recipes. If my wife ever wants a special dish for a special occasion I know which cookbook she goes for.

One day, while I was doing a word search for 'charcoal' on the Internet, Barbara's name came up. I never really knew her story and certainly I knew nothing about her charcoal connection. But, I called her, and she was every bit as friendly on the phone as she is on the screen. So I asked, "What is your story Barbara?"

Barbara explained that she had had Crohn's disease for twenty-three years. Crohn's disease causes a myriad of problems. She suffered with ulcerations of the large intestine, blood-loss anemia, malabsorption, chronic diarrhea, incontinence, and a rectal/vaginal fistula. Those that suffer usually suffer silently. Crohn's is an embarrassing disease, forcing its victims never to stray far from a restroom.

Barbara told me:

> "Six years ago I decided to change my lifestyle and my eating habits to control my Crohn's. God has blessed my efforts. Our miraculous bodies have been designed to be healthy when we put the right things in them. I have occasionally had bowel flare-ups, but can kick one out in twenty-four hours or less. For the past six years I have controlled my Crohn's disease drug free!
>
> "There is HOPE! My first attempts to eat a healthier diet were unsuccessful because most of the health food I tried tasted bland, coarse and most unpalatable. It has become my passion to create healthy food that tastes GREAT!"

If you want to learn more details about Barbara's story and how she keeps fit and active, you will find her web-address in this footnote.[3]

What she doesn't mention on her site is that with her initial doctor-supervised three-week juice fast, she also took charcoal. Barbara told me, "Along with the juice, I took 2/3 of a Dixie cup of activated charcoal powder mixed with olive oil. I ate this with a spoon every day for a month. The others [program participants] chuckled at me, and I had to laugh too. But the mixture helped to relieve gas, as well as aid in detoxifying my body of drugs and other intestinal debris."

People have written to Barbara and asked her about problems associated with Crohn's and other bowel diseases. So she has posted some suggestions on her web page that have helped her. She has given me permission to include her section on charcoal. Here is what she has to say:

"Some of you have written and said, "Help! What do I do about bad gas problems?" It is a problem and can be very embarrassing! I'm sure you all have at least a couple of stories you wish you could forget. I do too.

"Understanding why you have gas is half the problem. … There is a reason for it. Symptoms are not what need to be treated. The causes need to be removed! Don't you see that is exactly the reason some of us never get off medication for our entire lives? We have unfortunately become a society of symptom treaters. It's expensive, too! Don't get me wrong; I think there is a place for drugs. I just feel that it should be the exception instead of the rule!

"So what can you do? To treat the symptom at home and also receive some great healthful benefits, I use activated charcoal capsules. I buy them at a health food store or a local Co-op. When I have bad gas, I'll take six to eight capsules, three or four times a day, or whenever I am suffering from gas pains. Because it will remove and adsorb even the medication in your system, take it either one hour before medication or two hours after, just to be on the safe side.

"One tablespoon of activated charcoal powder can adsorb the amount of bacteria that would completely cover a tennis court. I also take four capsules if I eat something that gives me indigestion or stomach pain. I used it once in the middle of a store when my son consumed an unknown amount of liquid soap. We always carry a bottle in our glove box and in my purse for emergencies. For children who can't swallow pills, the capsules can be opened into a small amount of water, stirred and sipped with a straw. It doesn't have any taste. Don't try to put the powder on your tongue and swallow with water. It is so light that when you take a breath, you'll be sorry!"

Thanks Barbara for that motherly advice.

Night Call

Dr. Mary Ann, MD, has been a champion of a holistic approach to medicine for many years. She travels and lectures extensively. When she is not in her office she may be literally anywhere in this world. When she is home, she is sometimes called on to help other mothers in her nearby Amish community. She has given me permission to include the following case of cellulitis, which appears as the preface to her book, *God's Healing Way*:

"One cold January evening in Minnesota, my husband and I were settled in by our cozy wood stove when the phone rang. The call was from one of our Amish neighbors. Their six-year-old daughter had developed a skin infection, which had rapidly spread from the ankles up to the knees within the past twenty-four hours. The parents had tried a few home remedies; but as the short winter day turned into dusk, their hopes faded into despair. Would I please come to see what else could be done? Quickly, I packed my medical bag with poultice materials and herbal teas. Cautiously driving over the narrow snow-covered gravel road, I headed for their old weathered farmhouse.

"I found the little girl sleeping on a small cot in the middle of the dimly lit room. A worn blanket covered her shoulders leaving both lower legs exposed, as even the weight of a thin sheet on the sore limbs would have been unbearable. The little legs were swollen to nearly twice their normal size. Clear fluid was seeping through the pores of the taut, reddened skin. I stood there for a moment, assessing the situation—the exhausted pain-weary child, the anxious faces of the parents, and the solemn siblings hovering around the small quiet form. I silently sent up an urgent request for heavenly wisdom to meet this challenging situation. Then we went to work!

"The parents were instructed to fill two large buckets, one with hot water, with charcoal powder mixed in, and the other with cold water. The infected legs and feet were to be immersed alternately in hot then cold water for a total of seven changes. This contrast bath was to be given four times during the day. After each water treatment, a charcoal or herbal poultice was to be applied to the infected area. We prepared garlic and other infection-fighting herbal teas to drink throughout the day. I also prescribed plenty of pure water and a nutritious diet—free of sugar, grease, and lard.

"When I left the home later that night, the house seemed warmer and brighter. The family was filled with new hope and courage. When I returned the next morning, the father and mother happily reported that the pain in their daughter's infected legs had definitely diminished. The family members faithfully gave water treatments, applied poultices, prepared teas, and strictly adhered to the dietary plan. The pain, redness, and swelling gradually disappeared without a single visit to the doctor's office. This household was truly grateful for God's wonderfully simple healing ways!"

Some say, "Home is wherever you hang your apron." If you are sure to stock charcoal it won't be long before you will find charcoal is "at home and on the job" in your family as well. *CR*

---

[1] Notice this ad: "Developed for astronauts, used in gas masks during the Gulf War, and now available in a seat cushion near you—it's the odor-grabbing technology of Brand X. Formerly called the TooT TrappeR. You no longer have to be embarrassed by the untimely passing of intestinal gas among your friends or family. Product X has been designed and tested to absorb the odor and sound of flatulence. Malodorous gas is a naturally occurring event, obviously affecting some people more than others. Yet, clinical studies show that the average person produces one to three pints of gas and passes gas fourteen times a day."

[2] Many companies manufacture activated charcoals having different adsorptive capacities. The source materials and procedures give each charcoal a distinct pore diameter and internal volume that determine its adsorption potential. The U.S.P. (United States Pharmacopoeia) standard for activated charcoal has set an internal surface area of 1000 m2/g (square meters per gram). Recently, companies have begun manufacturing super activated charcoals, with up to 3 times the internal surface area per gram giving far greater adsorption power than standard activated charcoal. For example, Super Char (available from Gulf Western) has an internal surface area of 3000 m2/g. Actidose Aqua (Paddock) has a surface area of 1600 m2/g. In this book, all the recommended dosages for activated charcoal are based on the standard of 1000 m2/g. If you can obtain a super activated charcoal, use proportionately less.

[3] http://www.tasteofhealth.net/letter.htm

Chapter 10 Salting the Oats

Today, if someone is curious about some thing, one has only to slide up to a computer, type in a word or phrase, hit the enter key, and up comes a world of information. This is especially true for matters of health. There is all manner of material on any given subject. When it comes to understanding what constitutes good health, what is bad health, and what defines and identifies certain diseases, there seems to be a general consensus. But, when the matter of treatment is discussed, then the researcher is often overwhelmed with the varied and often widely differing strategies. More confused than when they began their research, some people head for the nearest drug store and purchase something they remember seeing on TV. That, or they meet someone on the way who mentions a remedy they or an acquaintance has used, and they decide to give it a try. The power of word of mouth should never be underestimated.

The Teacher

While at medical school, studying to become a naturopathic doctor, Dr. Dana, NMD, received a cat scratch on her foot. She paid no attention to it, but on the second day she noticed a red streak traveling up from her foot to her knee. The typical treatment would have been with antibiotics. Instead, Dr. Dana made a poultice with charcoal and water, and applied it directly over the scratch, and left it in place for a couple hours. When it dried out, she removed the poultice and was surprised to see the redness had receded half way down to the foot. She simply remoistened the poultice, and replaced it over the scratch. After a couple more hours, she removed it to find that the red streak had completely disappeared. She had no further problems.

After sharing her brief experience, Dr. Dana continued:

> "Between my first and second year of medical school, the grandfather of one of my classmates had been told by his doctor that his foot was so badly infected it needed to be amputated. He had already lost a toe, and did not want to go that route. I had the night care of this man and overnight I placed a charcoal poultice over the foot. We continued this practice for a month. We were rewarded for our efforts by seeing the foot turn from black back to a healthy pink. After the month, he was once more up and about.

We also gave him some simple hydrotherapy treatments, and instructed him on how to improve his dietary habits.

"I recommend to all my patients that they have a supply of charcoal in their emergency kits. For tooth abscesses, I take a tea bag and slit open the bottom. I then empty the contents, and replace them with charcoal powder from three to four capsules, and tape the bag closed. It is then moistened, and placed against the offending tooth."

The Teacher/Student

Dr. Wynn, MD, is one of the resident doctors at Uchee Pines Institute. Progressive-minded doctors are always ready to make room for innovative remedies. When I asked Dr. Wynn how he was introduced to charcoal he shared this story:

"The experience that really made charcoal real for me happened during my time working in the Amazon jungle in Ecuador. There you are surrounded by every conceivable parasite, but I had gotten off scot-free for some time. But one night we had our meal, and it just sat there in my stomach and didn't move. I wanted to vomit, but I couldn't. Next morning it was still there. The minister I was staying with pulled out some charcoal. Within a half hour of taking this wonder medicine, I was symptom free. I had heard how fast charcoal could work, but then I knew for myself."

Back in Alabama, Dr. Wynn relates:

"About ten years ago I was working outside wearing an old army surplus jacket. For some reason, I reached up with my arm and the coat briefly touched my eye. It felt as if I had slightly scratched the eye. It was irritating, but I thought it was nothing and would soon pass. But it didn't. After a couple days the eye was still bothering me. It felt like a grain of sand in my eye, but as hard as I looked, I could not see anything. I asked Dr. Don to look at it, but he couldn't see anything either. Finally I went to see an ophthalmologist. He searched, but could see nothing, and decided that I had a reactive inflammation on the eyeball. He suggested steroids and an analgesic. In fact, he put one drop of the painkiller on my eye, and it gave me the first relief I had had in days. But, within a half hour the eye was again bothering me. I asked the doctor what the danger was of not using the drugs. He said the eye could become permanently scarred.

"That afternoon, back at Uchee Pines, I was giving a class, and I mentioned my eye problem. One woman thought of a charcoal poultice for the eye, and immediately went and prepared some charcoal jelly with flax seed, enough to make several small poultices. But I didn't have time to put one on until that evening before going to bed. The next morning when I got up and removed the poultice, I felt excellent results—results that were just as good as with the analgesics that the doctor had given me!

"Recently our chief cook at the cafeteria accidentally hit his eye, leaving the white area bloodied. He was inclined to try some natural remedy so we again used charcoal, and the eye responded very nicely."

As you can see from the different stories, it is usually someone else that serves as the catalyst when it comes to trying charcoal.

Charcoal Internally

Salting the Oats

In trying to steer someone to better physical health, you often have to begin with the head, especially if it involves something new. For those of you who are already experienced with charcoal, and those of you who soon will be, there is a nice way of acquainting others with the unlikely benefactor of health—charcoal. We all know the old adage, "You can lead a mule to water, but you can't make him drink." BUT you can salt his oats!! In other words, we can add flavor, and create a thirst to try something new. Of course salt can also raise blood pressure, so sometimes "less is more". A friendly smile, a short convincing invitation will usually disarm most resistance. "Here try some of this."

In a large Finnish study, 102 patients were administered activated charcoal at home during the three-month research period. The study involved cases called to the Finnish Poison Information Center. The charcoal was most often administered with yogurt, crushed fruit, water, milk, or ice cream. Although the average dose in the study was 7.9 gms, patients took more charcoal when given with water than with dairy products (12.2 gms vs. 7.3 gms). In all but five cases in the Finnish study, there was success in administering activated charcoal at home.[1]

Unfortunately, a simple water and charcoal mixture can be messy and not very palatable. Dressing it up a bit with a thickener, a flavoring or sweetener can overcome these objections, without compromising charcoal's effectiveness.

Research studies showed that the addition of bentonite (clay) powder, with or without chocolate, significantly improved the palatability of a charcoal-and-water slurry. Bentonite acts as a thickening agent that reduces the gritty texture and improves the taste without reducing the efficacy of charcoal. The addition of chocolate further enhances the palatability of charcoal in research studies and in everyday experience.[23]

Salting the Oats

"Salting the oats", that is, making something that is rather unpalatable into something interesting, may mean dressing up the name and/or the product. Dr. Agatha Thrash suggests introducing it to squeamish children as 'chalk-a-sorb'—sounds remotely like chocolate. For the skeptic you may, like one mother, suggest "a little bit of black magic" to disarm them. Another mother uses the name "Jungle Juice". But when it comes to the actual product, it seems the imagination will fit the situation.

One large distributor of medicinal grade charcoal, recognizing charcoal's un-appeal, simply suggests that parents prevent children from seeing the black mixture as they drink it. Here are a few of the ingenious field-tested charcoal repackaging strategies: mash it together with a banana; put it in an empty chocolate milk container with a straw; mix it with peanut butter; or, give charcoal along with a dark, fizzy carbonated drink. Others suggest using cherry flavored or dark colored beverages. Since ice cream is often laden with artificial chemicals that compete for adsorption, it is not recommended as a disguise. One medical journal reports that mixing charcoal with an acidic juice (e.g. tomato or orange juice) makes the treatment for acetaminophen ingestion more palatable for children and does not impair effectiveness. I like the idea of charcoal Oreo cookies, but the idea is not new. During the early 1900s one home product was marketed as "Bragg's Charcoal Biscuits." Another old method to make charcoal more appetizing was to sprinkle it on buttered bread in the form of sandwiches.

While some children will follow adult direction without much fuss, others need coaxing. Whatever is compromised by mixing charcoal in fruit juice, chocolate syrup, jam, or honey to make it easier to swallow, will be rewarded by getting it to the troubled stomach sooner rather than later. Powdered charcoal reaches its maximum rate of adsorption in the stomach within one minute. So, in the case of poisoning, the sooner it is given the better the chances of a successful treatment. The dose can be repeated every four hours, or until charcoal appears in the stool.

If a mixed version of charcoal is better accepted, then simply give more charcoal to make up the difference in dosage. Ingesting high dosages does not interfere with sleep, appetite, or the sense of well-being. You can also put charcoal powder into empty gelatin or vege-capsules. These can be obtained at health food stores. Last but not least, you can, as some have already done, salt the oatmeal with charcoal—at least it works with livestock.

By Example

Remember, if children see a parent taking charcoal for their sore tummy, kids will be far more likely to take it too. Here is a case in point. Pamela had six girls.

The younger ones grew up seeing mom use charcoal regularly and hearing about its wonders. They also loved to go survival camping in the mountains. So, when their mother insisted that they take a survival kit along for snakes, etc., her daughters responded, "Look, there is a whole mountain of charcoal up there (there were many tree stumps left over from a controlled burn) and there is a pure stream of water. Why do we need an emergency kit?"

As for the run of everyday ills and hurts, eventually the whole family will be satisfied by just stirring the recommended dose of charcoal powder into a glass of warm water. It seems that the best time to take charcoal is between meals and when it does not conflict with taking drugs.

Charcoal Externally

Poultices

After diffusing possible objections to taking charcoal internally (many, after a brief introduction, are more than ready to try charcoal), there is the matter of using it externally. The rest of this chapter focuses on the external applications of charcoal. First, on how to use charcoal externally, then on what conditions are best suited to external use.

For the novice, charcoal can be as messy as making bread for the first time. In its powder form, it requires some care in handling. Experience has shown that it works better to add it to water, rather than the reverse. I also find that it mixes easier in warmer water. When it comes to messy, the difference between flour and charcoal is charcoal's tenacious ability to hold onto things. The fact that it is black means once it stains fabric it is very difficult to remove. So, when making a poultice, plan on using paper towels or old cotton sheets. But, like eating wholesome sweet bread, the end result of taking charcoal is a sweet stomach, sweet breath, and a renewed healthy disposition.

Before getting into charcoal poultices, let me say that the experts (those who actually use poultices with confidence) do not all do it the same way. There are no magic formulas to applying charcoal, but there are some easy guidelines. Because emergencies are most always inconvenient, and come at the most inopportune times, one has to be prepared to innovate. But, most any home already has many of the items you will need for giving a simple treatment, including making a poultice. Having your supplies all in one place, ready for quick use, is the first step to successful home remedies. For a start, purchase a large plastic container to store your materials in. Why not put charcoal on your shopping list right now.

In the section **Emergency First Aid** at the end of the book, you will find two basic formulas for making a charcoal poultice, one a "Jelly Poultice" with a thickener and a "Plain Poultice" without. The stories that make mention of a poultice, use one of these two methods, depending on preference and the availability of materials.

Charcoal Bath

For external applications, charcoal can be applied as a poultice or used in a bath. The charcoal bath is useful for treating large body areas. It's quite simple. For example, the feet. Fill a container big enough to soak the feet up above the ankles. Add warm/hot water, a couple tablespoons of charcoal powder, and enjoy the soothing relief. For a full body bath, just add one or two cups of charcoal powder to a tub of warm/hot water. The entire body may be immersed in the charcoal water. To finish, simply rinse off well under a shower.

Years ago when she had the luxury of taking a two-hour bath, Pamela (whom we met earlier) would put a cup of charcoal powder in the tub, pull a lap board up to her chest for some privacy (she had six daughters), and just soak up the heat while the charcoal soaked out the toxins. "The girls would read a story to me, and then help their hot wet noodle up to bed. By the next morning, I would be over whatever had been dragging me down."

Pamela's first introduction to charcoal's remedial properties came at a health seminar, but she didn't start using it personally until after she had applied it as a poultice to her injured horse. (Chapter 17) As she told me over the phone, "Well you know, you can hear a lot about charcoal, but you have to experience it. Then you never turn back."

Amputee

Pamela shared three remarkable experiences with me. I have included the other two elsewhere. The following is one she will never forget. You will also notice another vital ingredient to successful natural remedies.

It was several years ago that Pamela went over to the Philippines to help with public meetings promoting health:

> "Groups of young people were going door-to-door inviting people to the program. I would ask if anyone in the home was sick and needed help. But no one responded. But there was one young person in the group who looked in need. He was missing one arm, and his other wrist was bandaged. He had been with a group of young people climbing in the mountains, and somehow had grabbed an electrified line. The burns were so severe they had had to amputate the one arm. Now the doctors were going to decide whether or not they should remove the other. It had not healed and had become infected. Because of the circumstances, I offered to treat his arm. He was

unsure whether he should, so he asked his mother. She was so thrilled. She had been praying for something to save her son's arm.

"When I asked about getting some activated charcoal, she said she would have some for me within two weeks. I told them that I only had ten more days before I would be leaving. Instead I told her, 'Bring me some charcoal chunks, a bag, a hammer, and a boy to pound the chunks into a powder.' With some of the sifted powder, and a little water, we made a thick slurry. We washed and ironed some old sheets to make bandages. The boy had an inflamed surgical scar from his elbow down to his wrist. The doctors had tried to surgically remove the spreading infection. But it was his wrist that I focused on. There was an infected patch around his wrist about an inch and a half wide. The area had a putrid smell and was very painful.

"I applied the charcoal directly to the area and covered it. He returned that afternoon and told me that for the first time, the pain had stopped. The next morning when I removed the bandages there was no smell. The area showed new pink tissue. But the skin over the surgical scar also sloughed off, and you could see the suture marks. We then put a poultice up the entire length of the arm. The entire arm was well on its way to being completely healed by the time I left. The family was so grateful. The boy had told me, 'If I lose my arm I will end up a beggar.'"

No wonder Pamela will never forget that experience! But, the other vital ingredient I alluded to at the beginning of this story is a willingness to try a simple natural alternative to powerful drugs even when a positive outcome seems more than remote. Some will ask, was Pamela qualified? Yes, first and foremost, she cared enough to offer to help for free. Some licensed practitioners seem to act as if they would rather this young man would have lost his arm, than that an unlicensed, "unqualified" layperson was instrumental in effecting a rapid, complete and trouble-free recovery.

This next story is similar to Pamela's in that the charcoal that was used was not the more adsorbent activated grade, it was crudely ground, was applied as a poultice directly to the infected area, and the results were immediate and profound. While doing volunteer work in South Africa, Sarah met a man with a large boil on his wrist. His wrist was so swollen he could not wear his wristwatch. Doctors had tried antibiotics, but, as is the case more and more, there had been no appreciable improvement. So Sarah went to a nearby fire pit and collected some charcoal, pulverized it into a powder and mixed it with some olive oil. This she placed over the boil as a poultice and left it on over night. The boil burst allowing the pus to drain. Sarah wrote, "He continued the poultices, and by the third day his wrist was completely healed." Sarah now works at Wildwood Hospital, Georgia.

Over and over again, charcoal has proven itself to be a potent treatment for an amazing range of health problems. In many cases the simple remedies of nature, when used properly, work more effectively than some of the high-tech, so-called wonders of modern medicine. Charcoal has saved many lives, both in cases where medical help was not available and when it was.

Dr. Marjorie Baldwin, of Wildwood Hospital, Georgia, has often used activated charcoal to promote healing. She says that any inflammation, any area that is red, painful, swollen, and hot, responds to charcoal. She applies charcoal as a poultice if the inflammation is on the outside of the body, or gives it by mouth if the inflammation is in the digestive tract. Dr. Baldwin tells of a young woman with diabetes, who had developed ulcers on her feet and was threatened with gangrene. Her husband had frantically taken her to a very well known medical clinic in the Midwest. When the feet did not respond to careful treatment including antibiotics, she was told that she would need double amputation. She absolutely refused amputation, saying, she would rather die than live without her feet. Instead they went to Wildwood Hospital and Dr. Baldwin took charge of her case.

She writes in her e-mail:

> "I had the severely damaged feet put into separate double plastic bags partially filled with a slurry of charcoal and water at body temperature. The baggies were loosely tied above the ankles and a pillow was used to prop up the knees to keep the charcoal around her feet and from running out. When we completed applying the poultices, I and my helpers gathered around our patient and asked the Lord to bless the treatments and to guide our minds as to what more we could do."

Except for a brief period each morning when her feet received a sunbath and were allowed to dry, they were kept in the bags around the clock, with the charcoal mixture being changed four times a day. Her diet was also supplemented with extra Vitamin A in the form of a daily glass of carrot juice. This program restored her health, she was dismissed from the hospital, and she walked out on both feet.

Dr. Baldwin shared one other story about an eleven-year-old boy named Jimmy who was the unfortunate victim of a cigarette and a can of gasoline, leaving him severely burned on the backs of his legs: "He was brought to our hospital, and immediately laid face down on an examining table. In places his trousers were burned to his skin, and he was shrieking with pain. Immediately upon his arrival, I commandeered a crew of male nurses to prepare a large quantity of charcoal slurry. As I cleared his legs of all possible foreign matter, the crew smeared this thick paste of charcoal slurry on a piece of cotton flannel that would reach from side to side and waist to ankles. Then we quickly flipped this wet poultice, charcoal side down, over the burned area. Instantly the shrieking ceased, and the exhausted boy heaved a big sigh of relief. The cold charcoal had completely relieved his pain. All that Jimmy had was charcoal. He was spared his limbs to walk again."

This case happened before the advent of cold therapy for burns. After the initial healing, Jimmy received skin grafts to complete his recovery.

What Dr. Baldwin used regularly in her practice, other progressive physicians are also discovering. *The Lancet*, the prestigious British medical journal, describes the use of charcoal compresses to speed the healing of wounds and to eliminate their odors.[4] This article tells about the amazing ability of human skin to allow the transfer of liquids, gasses and even micro-particles through its permeable membrane and pores. By the application of moist activated charcoal compresses and poultices, bacteria and poisons are drawn through the skin and into the charcoal. Poultices must be kept moist and warm to allow this healing process to take place.

In following up the *Lancet* article, I noticed that this new trend of using charcoal impregnated dressings is catching on in the U.S., Canada, and elsewhere. The Australian Government Dept. of Health and Aging lists in its national dispensary an activated charcoal bandage as well as a charcoal foam dressing for managing foul smelling wounds.

Kidney Failure

If charcoal applied externally can produce such remarkable results for surface wounds will it work for internal problems?

While working as a lifestyle counselor at Uchee Pines Institute, Joel was assigned to care for a twenty-six-year old man from New Jersey. Ustes was a welder by trade, and had become acutely ill from breathing the fumes while welding galvanized metal. These deadly fumes are known to destroy the nephrone cells in the kidneys, keeping them from properly filtering the blood. In his case, Ustes' blood work showed extremely elevated levels of waste products so he was scheduled to begin kidney dialysis in a couple of weeks. At this point he came to Uchee Pines and was put on a cleansing program. To relieve his kidneys of some of their workload, he was given a diet low in proteins. He was given a daily walking program out in the sunshine and fresh air. Of course there was time to rest, and to meditate upon the blessings of God, but his main treatment consisted of a daily charcoal bath. This was supplemented each night with a twelve-inch wide poultice applied around his trunk. Joel also noticed what I and others have detected—a distinct urine odor and yellowish stain on the poultice. How that happens is a mystery that science does not fully understand. The first and second weeks showed a steady drop in Uste's blood waste products. By the third week his blood functions had returned to near normal. When Ustes returned home he committed himself to staying with his new health program. Ten years later when Dr. Calvin Thrash contacted him, he reported that he had continued on the program and had never needed dialysis.

Different Poultices

Poultices have been found to be effective for itchy skin, infections, gangrenous ulcers, insect bites and stings. When one considers that the skin is the largest organ of the body, it is reasonable to expect that poultices would have the profound effect they do. However, the poultice does need to be kept moist to be effective. Charcoal may be used by itself or with ground flaxseed or with clay added to the charcoal. Using equal parts of charcoal, clay (Bentonite, Aztec, or Montmorillite) and ground flaxseed makes a nice poultice that will, when covered with plastic, remain moist overnight. Once used, poultices should be discarded.

For inflammation of the eyes, apply a small 2 in. x 2 in. charcoal poultice over one eyelid for eight to ten hours. It should then be removed and another compress used on the opposite eyelid. You can use a ski cap or surgical tape to hold the poultice on the eyelid. Those who are writers and end up with tired eyes will find this simple remedy helpful.

A little charcoal paste or a moistened charcoal tablet applied to a band-aid works well for insect bites.

As we have seen in different cases, patients with a diseased liver or kidney can be treated at home with a large compress over the mid-back or over the abdomen. To hold the poultice in place, plastic stretch wrap can be wound around the person's stomach. The best time to do this is just before going to bed. In the morning before bathing, the poultice can be taken off.

Making Your Own Charcoal.

As skepticism in charcoal's curative powers gives way to wonder, we are naturally drawn back to how it is made. It seems the mystery of charcoal's nature begins right in the making of it:

> "We want ours to burn good and slow," said Young Billy. "If he burns fast he leaves nowt but ash. The slower the fire the better the charcoal."
>
> Susan was watching carefully. "Why doesn't it go out?" she asked.
>
> "Got too good a hold," said Young Billy. "Once he's got a good hold you can cover a fire up and the better you cover him the hotter he is and the slower he burns. But if you let him have plenty of air there's no holding him."
>
> —Arthur Ransome
> "Swallows & Amazons" 1930

When making your own charcoal, choose a species of hardwood. Softwood will work, but you will get very little for your effort. This would not include used building materials that may have been painted. If you have a fireplace or wood stove, after the fire has gone out overnight, the leftover pieces of charcoal can be used. But the ultimate in making your own charcoal begins with a wood fire out-of-doors.

You can arrange the wood above ground or in a shallow hole. After the wood is burning brightly, it can be covered with a large piece of tin, and dirt piled over the tin to make a dome to exclude air. As the heat continues to burn the wood without oxygen, the soft parts of the wood are burned out, and the hard parts remain, making a good charcoal. The charred parts of wood should then be put in a cloth bag and pounded to coarse granules. After it is reduced to chunks ranging in size from small peanuts to rice grains, put the charcoal into a blender and pulverize it to as fine a powder as possible.[5]

Making activated charcoal at home is neither practical nor safe. Activated charcoal is clearly superior to homemade charcoal but do not be put off from using the simplest form of charcoal just because you do not have the more adsorptive activated grades. As we have seen, when no commercial grades of activated charcoal were available, charcoal still worked even in its more basic form.

Repeat Learners

We began this chapter showing how those who teach health often find themselves as repeat learners. Having discovered for themselves some new remedy, they look for ways it may benefit others. In turn those who are helped will find others.

Early in the book, I mentioned my experience of grave sickness on the remote Pacific island of Abemama. It was not my last painful experience there. I am a slow learner. I'm embarrassed to confess this, but no sooner had my foot and leg infections begun to clear up, than I stepped on a larrrrge nail. You would think that I would have been extra careful. The foot began to swell immediately, and once more it was back to charcoal. Once again I added charcoal to a hot foot bath. I was grateful that the wound did not become infected, but my immune system was still low and so it took longer to heal than it normally would have.

When my friend Wayne arrived to help with our garden training project, I noticed that no matter how hot his feet got, he always wore adequate footwear. Hot or not, I finally decided to try his natural remedy and wear proper footwear and my feet quickly finished healing. By then I was also far more aware of, and sympathetic to, what others were suffering.

About that time I noticed that Aaba, one of my students, had an open ulcer on his arm. He acted brave but I knew it must have been terribly painful. One evening after a school program, I mixed up some charcoal paste, took some scissors and clean cloth, and Wayne and I went to visit Aaba in the boys' dorm. In the large dimly lit foyer, pressed in with more than a hundred curious young men, Wayne helped as I applied the poultice and wound his arm with cloth. With everyone trying to get close to see what was going on, it was near bedlam. When we were done I invited Aaba to bow his head as we asked God to bless the simple poultice. The room became instantly still. When I finished there was a unanimous youthful shout, "AMEN!"

As we were leaving, two other young fellows pulled a very reluctant friend over to see us. He too had an angry-looking open ulcer on his arm. We had just enough charcoal for another poultice. The next day, as they marched out of the chapel service, there were two very happy looking believers in charcoal. Wayne accepted the job of changing the poultices every day until they were fully healed. Teaching, learning and teaching again is a process we should never tire of. Along with the sacrifices it offers great rewards.

Many other local people suffered from similar infections that went untreated and left massive scars. I began to understand how important it was to treat infections vigorously at the first sign. Of course it helps to believe in what you are doing. Because I had already seen charcoal poultices work, it was a lot easier for me to apply them to those students and have faith that they would work for them as well. Your faith in charcoal will grow too as you are thrown into circumstances where you feel inadequate, only to find charcoal was up to the job. In a day of litigations over drug induced complications and death, charcoal remains a quiet, harmless, tireless worker in the classroom, in the hospital, in far away places, and right in the home. CR

---

[1] Lamminpaa A, Vilska J, Hoppu K. Medical charcoal for a child's poisoning at home: availability and success of administration. *Human Experimental Toxicology*, 12:29-32,1993

[2] Gwilt PR, Perrier D, "Influence of thickening agents on the antidotal efficacy of activated charcoal", *Clinical Toxicology*, 19(8):89-92 1976

[3] Topuzov EG; Beliakov NA; Malachev MM; Umerov AK; Solomennikov AV; Gritsenko IV; Kokaia AA, "Use of enterosorption in biliary tract cancers complicated with mechanical jaundice". *Vopr Onkol*, 42(2):100-3 1996

[4] *The Lancet*, Sept 13, 1980

[5] Thrash, *Home Remedies*, p.144

Chapter 11 Spiders and Company

When we think of "Nature", the picture that comes to mind is often colored by our experiences. It might be beautiful scenery, peaceful retreats, exciting adventures, and intriguing mysteries. Or it may be spiders, wasps, bees, hornets, ants, scorpions, snakes, and poison ivy. How often childhood memories of being out in the country are "bitten" with painful recollections. But they need not end on a sad note. Charcoal is ready to take the sting out of our painful experiences, **IF** we have it ready at hand.

I grew up in rural Florida, with a healthy fear of black widow spiders. I was never bitten by one, and judging from all the scare stories I heard, I never wanted to be. So when I started asking for stories about insects to include in this book, black widows were the first ones to come to mind.

Spiders

Dr. Dana, NMD, told me over the phone:

> "I have never actually treated anyone with a bite from a black widow, but have treated several brown recluse spider bites. These are far more difficult to treat. I have seen people who required cosmetic surgery to repair the extensive tissue damage, and some people have recurring symptoms years later."

She went on to tell of an encounter a friend had with a brown recluse:

> "While teaching a course in Scottsdale I noticed a small ulceration on my friend's leg. Maxine had not noticed it, but by the next morning it had gone from less than dime-size to quarter-size and had become angry and inflamed. A large open ulcer had developed, and she had enlarged lymph nodes in the groin. I immediately prepared a poultice, and placed it over the open ulcer. The poultice was replaced several times daily over the course of a week. Maxine also took a couple of herbal supplements and charcoal internally. By the second week there was no evidence of a bite at all."

The brown recluse spider produces a bite that gives little or no pain at first, but is extremely toxic. This creature, with the fiddle design on its head, is more to be feared than the black widow spider—which is more easily identified. Within twenty-four hours a purplish-red blister develops at the site of the bite, and extensive tissue death occurs beneath the bite. This produces a very deep and angry ulceration that may extend down to the bone. The condition often lasts for weeks or months, and typically leaves a deep puckered scar. That is, if amputation does not become necessary. There is no antidote and no anti-venom. So, in hopes of physically removing all of the poison, the treatment often resorted to is that of wide surgical excision of any flesh containing venom.

However, it has been found that a totally benign treatment for brown recluse spider bite is a compress of powdered charcoal. This should be applied as soon as possible after the bite happens, preferably during the first twenty-four hours. For the first eight hours, change the compress about every thirty minutes. On the second day, the time interval for changing the poultices or compresses can be lengthened to two hours, and then to four.

In their book *Rx: Charcoal*, the Doctors Thrash include three experiences with brown recluse bites:

> "Our third case was a sixty-year-old man who called us from a Veterans Administration (VA) hospital. He had been bitten by a brown recluse spider two days before, having seen the spider and recognized it. He now had a large blood blister on his thumb just at the base, about the size of a dollar coin. There was extensive swelling of the entire thumb and forefinger, with purplish discoloration of his whole hand. In an attempt to save his hand, the VA surgeons offered him as their treatment of choice, the removal of his thumb and the fleshy mound of muscle and bone at its base right back to the wrist.
>
> "At that point he called us. We advised charcoal compresses changed every thirty minutes for the remaining eight hours of the day, and every two hours during the night. Three months later, he drove over to show us the results of following our advice. He had a shallow, elastic scar which did not interfere at all with the movement of his thumb! The very fact he had a thumb at all, would have caused the VA doctors to marvel."[1]

The Thrashes also tell the story of an anesthetist who attended one of their heath seminars. On hearing one of their case reports about a patient's recovery from a brown recluse bite, he related his experience from the previous day. He had had to anesthetize a lady for the second time in two months so that a more extensive amputation could be done on her foot. She had been bitten by a brown recluse spider. The first amputation had not removed all of the damaged tissue, and the foot had failed to heal.[2]

Dr. Churney, MD, relates his experience with a recluse spider bite:

> "Several years ago while working at Wildwood Hospital in Georgia, I attended a man from Pennsylvania who came in with a brown recluse spider bite on his leg. He had gone to emergency in Pennsylvania and had been told they would excise the muscle tissue around the bite. He left and drove down to Wildwood. He was treated with a charcoal and flax seed poultice over the bite. We alternated the poultice application with an improvised oxygen tent over the leg. We directed the oxygen right at the bite area. The poultice was changed twice daily. By the third day his leg had returned to normal."

How fortunate it was that this man knew of this reputable hospital. But not everyone has access to a modern hospital emergency ward, especially one that, in many cases, encourages the use of simple remedies as a first line of treatment.

If someone wanted to know the strengths of some simple remedy, it would seem reasonable to inquire of those who regularly use that remedy, and know firsthand how well it works. Their information will be more credible than those "armchair" experts who can supply facts, but have no personal experience themselves.

Emily has gained an experience with charcoal in a relatively short time and is certainly another one who qualifies to tell the merits of charcoal. She and her young family live out in the Michigan countryside and have a problem with spiders in their two-year old yard. She writes:

> "Because we have young children, and because I am sensitive to chemicals, we do not like to spray insecticide. Our spiders are not small, innocent ones. We have black widow and brown recluse on our property. Even in the summer, I ask the children to wear jeans and sneakers when they are out.
>
> "One Sunday evening last summer, I decided to cut my son's hair out in the backyard. Not wanting to get hair on my clothes, I wore shorts and sneakers. My son had jeans, sneakers, and no shirt. He was fortunate, but I was not. After finishing his hair, I moved his chair to clean it off, and a brown recluse jumped on my left leg. It bit me, and jumped away. This was my first spider bite encounter, and I did not really know what type of spider it was until I had confirmed it with several sources. I am a firm believer in medicinal charcoal, so I immediately put a small poultice on the bite. I kept it on until the next day, Monday, not knowing how serious this bite was.
>
> "In the afternoon the bite still hurt a little, but I just put a bandage on it and went out to a ball game with my son. When we got to the ballpark an hour later, my leg was hurting quite badly. I did not want to ruin my night out with him, so I tried to ignore it. By the end of game, I had a very dark red line running just about an inch up my leg from the bite site. When we

finally made it home, I was in so much pain that I was surprised I had not stopped into the nearest emergency room. I quickly made up another, bigger, charcoal poultice and applied it. Very soon the pain began to ease. I changed the dressing twice a day for the next four days. Then I left it off to see if it felt better and it did. I will gladly share the cure with anyone else who needs it."

As we will see later, not only did Emily share her story, but she also shared her charcoal and time to help others.

Bees and Yellow Jackets

Like Emily, Erna and Kelly, who we met earlier, also live in the country. They have become quite knowledgeable in the use of charcoal. Their experiences are typical of those who know the healing wonders of charcoal firsthand. Erna writes of one of their earlier experiences: "Sarah was about eight when she was unfortunate enough to receive two bee stings at the same time. We were traveling, and I didn't happen to have my charcoal with me, so we didn't do anything at the time. In a couple of hours we were home, and she was no longer complaining. However, by bedtime she said she didn't feel very good and her bones ached. I took her temperature. It was somewhat elevated but I don't remember how much. It took a few minutes to figure out that she was having a reaction to the stings, but when I realized what it was, I put charcoal paste (just charcoal powder and a bit of water) and a band-aid on each of the sting sites. I also gave her one heaping teaspoon of powder in a glass of water to drink. She said her prayers and went to bed. When I checked on her later, her temperature was down, and she was feeling normal again."

Living in the country does expose one to more of nature's insect traffic. Living in the country also tends to make people a little more independent. Because of time and distance to city services people often learn to wait and innovate. Natural remedies like charcoal fit very well with country living. But, come a long weekend, and countless thousands of city dwellers leave the city limits behind to visit their country cousins and enjoy the wide open spaces for themselves. Natural remedies like charcoal fit very well with city campers too.

Pauline and Leo are a case in point. Earlier, I mentioned Pauline's use of charcoal in her hospital emergency room setting. However, many emergencies are better dealt with right where they happen. Not long ago, Leo and Pauline purchased a property, at some distance from the city, and began clearing it for their new home. Pauline writes: "This past Monday Joshua, our six-year old, got stung on his ear by a honeybee. He screamed and screamed. We knew that he has been very sensitive to bug bites, and were concerned that he might have an anaphylactic reaction. While I tried to calm him, Leo went over to the fire pit where we had had a recent fire, and he got a piece of charcoal. He quickly ground it up and made it pasty with some water. We plastered that on Joshua's ear. He was still screaming up to that point, but soon after the charcoal was applied, he calmed down. The swelling stopped and he had no other bad affects after that."

It is understandable that working in an ER, where medicinal charcoal is always available, would make it easy for Pauline to forget to take some with her camping. However, in an emergency, the primitive form of charcoal is still a very good first aid.

I only met Jacob a couple of days ago. Because he is affiliated with a holistic health-conditioning center I once visited outside New York City, and because it has become a habit, I asked him if he had any charcoal stories. He said this experience came immediately to mind: "When I was in Georgia, a four-month old girl was bitten on her hand by a yellow-jacket wasp. In just a matter of seconds, I saw the arm turn purple, beginning from her hand and reaching all the way up her arm. I quickly made and applied a charcoal poultice to her entire arm and hand, and I gave her a charcoal slurry to drink. Slowly the color retreated from the biceps downward, until within about thirty minutes the color of her whole arm was back to normal." No wonder he remembers the story!

Doctor Agatha Thrash (pathologist and, for forty-five years, Medical Examiner for the State of Georgia) knows how dangerous insects can be. She relates several experiences in this chapter. This one is also from her book:

> "A three-year-old girl was playing in her yard. Seeing a hole in the ground, her curiosity was piqued, and she stuck a stick into it. Out came a swarm of angry yellow jackets which immediately attacked her. Hearing her screams, her mother came running to the rescue. By that time the little girl was covered with the vicious insects. We later counted over fifty stings on her from the collarbones up. The frantic mother began to beat the wasps off, and in the process she got fifteen or twenty stings herself.
>
> "Hearing the mother's call for help, several people came running. They immediately took and placed the girl in a tub of cool water, covering everything except her nose and mouth, and stirred in several tablespoons of charcoal. They kept the girl in the tub for about thirty minutes. After cleaning her up, she seemed perfectly comfortable and was soon playing again. The mother, who had been too busy to care for her own stings, had marked swelling and pain that persisted for several days.
>
> "Mrs. T. had become extremely allergic to bee stings. With her last sting, she had nausea, weakness, faintness, and some wheezing, which had necessitated treatment at an emergency room. Her physician had warned her that the next attack could be fatal, and urged her to undergo a series of desensitization

shots. Unfortunately, before she could do so, she got another sting on her hand. Right away, the lady was pale, sweaty, weak, had a headache and nausea, and severe pain in her hand. She was beginning to wheeze.

> "We first rubbed the sting with a charcoal tablet wet with water as the very quickest way to apply charcoal, and immediately mixed some activated charcoal and water, and more completely covered the sting. Within two or three minutes of the charcoal application, she began to relax and feel better. A larger poultice was prepared to replace the emergency one, and it was changed at ten minute intervals for an hour. We gave her a tablespoon of charcoal by mouth in a glass of water. Then an interesting thing happened.
>
> "The patient felt perfectly well and took the poultice off believing herself to be entirely finished with the reaction. Within ten minutes, she was weak, sweaty, faint, and beginning to wheeze. The poultice was immediately re-applied, again with clearing of symptoms. After wearing the poultice all night, she was well. Although this lady usually experienced massive swelling after bee stings, she had no trace of swelling."[3]

As an ounce of prevention for the unforeseen, those who know they are always at risk from severe reactions to insect bites should always have charcoal with them. It's too late to go buy a fire extinguisher when your house is burning. Those who know also tell us that most people who die in fires had no emergency plans. Charcoal will be a part of any well-thought out emergency preparedness, whether one lives in the country or the city.

Poison Ivy

Several years ago, Esther went to Uchee Pines Institute in Alabama to attend their health program. While there, she was introduced to charcoal, and it made a lasting impression. Esther returned home to Georgia and started a medicinal charcoal business that continues to grow. When talking with Esther, I asked if she would share some of the stories she has gathered over the years. The one that caught my attention was about the farmer that came and bought a large supply of charcoal for his dairy cows. She can't remember the details, but maybe by the time we get to Chapter 17 it won't seem like such a mystery. Esther shared several stories including the three that follow.

Another painful experience for many people is contact with poison ivy or some other poisonous plant. Esther writes:

> "One man I dealt with had so severe a case of swelling due to poison ivy that he couldn't open either his fingers or his eyes. His ankles too were terribly swollen, and there was a rash wherever the poison ivy had touched.

He took eight charcoal tablets twice a day, and by the fifth day the swelling had completely gone."

Esther related this story from her friend Mr. Choi:

"After returning from picking some plants in the forest, one of the ladies that was with us developed inflammation and itchiness on both her arms. I gave her one large spoonful of charcoal powder and applied charcoal poultices on the affected areas. One hour after the application, she said that the area felt hot and painful. But the next day she wanted to apply more charcoal on her arms because the inflammation was gone. She applied charcoal poultices for a week and had no more problem."

Mr. Choi is quick to recommend charcoal because he knows personally how well it works:

"As I was taking care of the cattle on the farm, I noticed that my eyes were getting terribly itchy. My doctor prescribed some cream, but when I applied it to my eyes, they swelled and turned red. So I washed thoroughly around my eyes and applied a gauze poultice using flaxseed and charcoal powder. After a day the infection began to clear, and after three days of charcoal treatment the infection was completely gone."

It is a good idea when one realizes that they have brushed against some poisonous plant, to immediately wash the area well with soap and water. After this brief preparation, do as Mr. Choi did, apply a charcoal poultice and take charcoal internally. If you follow his routine, in a couple days you too may be able to wink with a smile.

Ants

Dr. Wynn, MD, understands how serious a small bug bite can be for some individuals. He told me about his wife's sensitivity to ants. "My wife and her sister have very strong reactions to ant bites. Her sister reacts so badly that one time we had to call 911 because she was having an anaphylactic reaction. When my wife had her next bite, she immediately took some charcoal orally, and then put some directly on the bite. She had no adverse reaction to the bite at all."

Aware of the dire results of some insect bites, Dr. Thrash takes no chances and makes sure that charcoal is never far away. She shared this personal experience:

"My little grandchild accidentally sat down on a hill of fire ants. Instantly hundreds of ants began biting her. She screamed hysterically from the intense pain. We grabbed her, stripped off her clothing, and ran for a bathtub. As it filled with water, I added charcoal. After being submerged in that charcoal bath for less than two minutes, she stopped crying. The charcoal neutralized the poison, and her pain was gone. Charcoal has amazing healing properties. In fact, if I were stranded on a desert island, and could take only one thing along to protect me from disease, infection, and injury, I would choose charcoal."

Now that is prevention in action. So for those of you who know you have severe sensitivities to insects or plants, it would be prudent to carry some charcoal with you on all your outdoor ventures. For those of you who are less sensitive, be sure to take some charcoal with you so you are prepared to help others, if need be. One never knows what might be lurking on a limb or under a leaf.

Scorpions

I met Sandra several years ago when I was working in Nepal. She travels extensively back and forth through Nepal and northern India. She gives health lectures, and teaches proper nutrition, hygiene, sanitation, and simple home remedies. Ignorance and superstition prevail throughout the area. She has had numerous experiences and shared these following stories in a newsletter. Sandra writes:

"We have taught the use of charcoal in the treatment of poisonous bites and stings. One man had been climbing a tree, cutting leaves for his animals, when a scorpion stung him. One thing that the witch doctors say is, that you should never kill the scorpion that stung you, or you will die and you should never drink water after being stung, or you will die. But this man killed the scorpion and then came quickly back to the village and met with a sister who knew how to use charcoal. She applied a charcoal poultice, and gave him some charcoal water to drink. Within an hour he was well.

"A couple of days later we were at a cottage meeting. A man entered and greeted everyone. He then put his hand up on a beam in the house, and was immediately stung by a huge white scorpion. This man had been questioning whether the witch doctor or God was truly stronger. Now he had a chance to see firsthand.

"Immediately his hand started to swell. He put a constricting bandage on his wrist, but the pain continued up his arm to his heart. He felt a burning sensation in his stomach, and he knew he would soon die. After killing the scorpion, this same sister, who helped the first man, quickly gathered some hot coals from the fire, cooled them, and then ground them to a powder. With the help of others, she made a charcoal poultice and put it on his hand. She then gave him charcoal water to drink. The man was sure he would die from drinking water. But with the help of charcoal and prayer, by the end of the meeting, an hour and a half later, he was fully recovered."

Whether you're bitten by ants or spiders, or stung by wasps, hornets, bumblebees, honeybees, or scorpions, charcoal will quickly relieve the pain … if you have some at hand to use.

Snakes

It seems right to have a healthy respect for things like scorpions, yellow jackets, and fire ants especially after you have been stung. However we all come alert at the mention of "snake!"

While on a hike, John, who was about fifteen at the time, encountered a small rattlesnake. When I called, he gave me the details:

> "I had been successful in picking one up earlier in the day. So, I proceeded to try again. Only I did not grab it close enough to the head. It was able to squirm around and bite me on the second knuckle of my index finger, but only with one fang. Instantly there was pain. Over an hour elapsed before I was able to get back to my father, who was a doctor. My arm was swollen tight to the elbow, and the pain was severe.
>
> "To allow for some bleeding, my father made an incision near the puncture. Then, from 3 p.m. till midnight, he applied a tourniquet, releasing it and tightening it as I showed signs of faintness. It was about two hours before we were able to apply any charcoal. My mother, an RN, made up a thick charcoal paste and smeared it over my hand. She put gauze over the paste, and then a plastic bag over the entire hand.
>
> "Later, they immersed my entire arm in cold water with pulverized charcoal. Mother also gave me Blue Kohosh tea to drink. By the next morning the swelling had almost disappeared and soon I was fully recovered."

Oh yes, John's father is the same doctor that would later introduce me to charcoal in Guatemala.

There are few places that are free from poisonous snakes, but there are some places where you think you should be safe.

Late one North Carolina summer evening while working in the family garden, Bob was cultivating around some potato plants when a half grown copperhead snake struck and bit him on his finger. Bob writes:

> "It felt like the sting of five to ten yellow jacket wasps all at once. The bite had the appearance of a typical bee sting. There was the typical white circle about a quarter of an inch in diameter, with a red dot in the middle. The swelling began immediately, and within two or three minutes my finger was swollen and getting stiff, with the swelling extending into the back

of my hand. Back at the house, I applied a rubber band tourniquet to the base of the bitten finger and washed the area with hydrogen peroxide. I next lanced the fang marks with a razor blade sterilized with alcohol, and squeezed the finger to aid bleeding."

Bob also drank about six ounces of charcoal water, and soaked the area in a basin of iced charcoal water as his wife Vicky prepared a charcoal poultice. This they applied to the wound making sure to keep the poultice moist. The poultice was made by stirring together enough charcoal and water to make about two quarts of liquidy paste. Laying out a paper towel, the paste was spread on it liberally. They then laid the hand and arm in the paste. The rest of the paste was spread evenly between the fingers and over the hand and arm. The edges of the paper towel were then pulled up to completely wrap around the charcoal. To slow blood circulation, the hand was also wrapped with plastic bags filled with ice. Together they then asked God to bless their efforts.

By midnight Bob's hand and arm up to his elbow were very swollen His hand was twice its normal size and the swelling was progressing up to his shoulder. Bob also noted that, "the actual pain from the bite only lasted until charcoal was applied and apart from some sensitivity to the pressure from swelling, I did not experience any severe discomfort."

For the next week Bob and Vicky applied poultices of charcoal or plantain, and ice bags. Along with herbal teas, Bob also took various supplements to promote healing. By Sunday morning, five and a half days later, Bob's hand and arm were almost back to normal. There was still some puffiness in the finger, hand and upper forearm, but this too passed. A month later the only lingering effect was a slight stiffness in the finger… and a more watchful step in the garden.

The Doctors Thrash have not had firsthand experience with treating a poisonous snakebite with charcoal. As such they are hesitant to suggest using charcoal exclusively in treating snakebite. But, judging from the personal experience of others they feel it should certainly be considered as a first aid treatment. They include this case they monitored over the phone:

"Our first case was reported to us by telephone. A couple living in a remote area of Arkansas, sixty miles down a winding road from the nearest hospital, called in great distress to report that their one-and-a-half-year old son had been bitten on the chest by a copperhead moccasin. This snake may cause death in a small child. The fang marks were surrounded by some swelling, and the child was in great pain.

"We instructed the couple to get to the emergency room as quickly as possible. We also told them to apply charcoal poultices one after the other, every ten minutes until they got there. We were greatly relieved when they called several hours later to report that the child was doing quite well.

"In fact, by the time they had gotten to the hospital, the child was asleep, the swelling had gone down and the pain had obviously stopped. The physician administered the antivenom to be on the safe side, but told them that he didn't really believe it was necessary. The child had no ill effects."[4]

Scientific experiments over many years attest to the effectiveness of charcoal as an antidote. In one experiment, one hundred times the lethal dose of Cobra venom was mixed with charcoal and injected into a laboratory animal. The animal was not harmed in the least.[5]

Nepal

In a laboratory, experimenting on an unfortunate animal is one thing. A day or more away from the nearest hospital, trying to save a person's limb or life is quite another.

After their first year in Nepal, Joel and his family moved out to where they would be working in the small village of Huwas. They stayed at the small school near the building site for their new home and clinic. Someone brought a man in to see them who had had a sore on his foot for many months. It was red, swollen, and the skin was sloughing off and very painful. On piecing the story together, Joel found out that the man had punctured his foot by stepping bare-foot on the bones of a snake lying on the path. If the foot wound resulted from snake venom there was no known remedy, and it was considered incurable. The man had gone to the distant hospital, and had tried everything from antibiotics to the local witch doctor. However, his foot did not improve.

So Joel explained to the man how to give himself hot and cold foot baths (Chapter 17), and made a charcoal poultice for him to apply for the night. The man returned the next day much encouraged, as the pain had subsided considerably. They gave another hot and cold treatment and changed the poultice. By the second week, apart from a scar, the foot was completely recovered.

Ragendra, one of Joel's young Nepalese translators, did not show much confidence in charcoal. While visiting with his family in nearby Walling, a woman in the town was bitten by a poisonous snake. She began wailing, sure that she was going to die. The pain was so bad that she did not sleep for twenty-four hours. Hearing of the woman's plight, Ragendra went to the water buffalo shed and collected some charcoal. He pulverized it and mixed up a poultice as he had seen Joel do, and he put it on her leg. Within fifteen minutes the woman fell asleep.

By then it was too late to take the bus to the hospital in Tansen. But the next morning he took her on the five-hour trip. It was several more hours before they finally got to see the doctor. When the doctor unwrapped the soiled bandages, he was surprised to see charcoal. "Who knew to put on charcoal?" The doctor's verdict? "It had most likely saved her life!"

We are all a bit like Ragendra in that we tend to think "real" diseases and "real" pain need something more uncommon than just charcoal. But, circumstances have a way of showing us just how "real" and powerful natural remedies can be, including charcoal.

India

For many years, Beth and John worked next door to Nepal in India, and saw suffering on all levels on a daily basis. They are confirmed users of charcoal, and are aware of its varied applications. India is a land of extremes, including emergencies. A bite from the Russel's viper is one such extreme emergency.

Before we hear Beth's story, let me give you a little background on the Russel's viper. Its venom is extremely lethal, destroying the walls of arteries and causing heavy internal hemorrhages. Victims often begin to cough blood. A bite can be fatal within forty-eight hours, if it is not treated with antivenom with an immediate positive effect. There was a time when this snake killed more people in the world than any other snake! The Indian cobra may be the most revered and respected snake of Sri Lanka, but the Russel's viper is probably the most feared and hated. Hated because of its deceptive nature, as it is often confused with a harmless snake.

Beth tells this remarkable story:

> "A woman was bitten by a Russel's viper. Her neighbor immediately pulverized some charcoal and, making a paste, plastered the area with charcoal. The woman was then rushed to a hospital, but there was no antivenom in the area. The woman suffered no complications! Because there were no side effects, her husband would not believe that the snake was poisonous. So he had an expert come. But the expert positively identified the snake to be venomous. The woman was convinced it was the early use of charcoal that had saved her."

Africa

Whether you are in America, Europe, Asia or Africa, there is little comfort when it comes to being bitten by a poisonous snake, unless…

Carl and Beverley are currently involved with development work in Uganda and southern Sudan. I happened to discover the following snake encounter posted in their web diary. They have allowed me to share it. Beverly writes:

"Sept. 3, 2004, Arua, Uganda—One of our Guards, Samuel, came to the house early this morning with the report that his wife had been bitten by a snake last evening. She had made it through the night, but the leg was swelling and painful. I brought her home and put charcoal poultices on it (used cornstarch to thicken with), covered that with a plastic bag, and changed it about every half hour. I also gave her two teaspoons of charcoal to drink every half hour for the first hour, then once an hour for the rest of the day. Within thirty minutes the pain had significantly reduced and within an hour she could bear a little weight on the leg. By evening she was walking with a slight limp. Only the toe that was bitten still remained painful. She will keep charcoal on it all night.

"Sept. 5 —Update. Samuel reports that his wife is fine. She was weak for a day or two, but was able to move about. Today she is back to normal."

The west side of Africa also has its share of poisonous snakes. Clyde and Cathy, RN, are directors of an international charitable organization and were on a supervisory trip to visit a small family clinic in Guinea, West Africa. While they were there, a man who had suffered a snakebite on his foot several months previously came to the clinic. Cathy writes:

"His foot was badly infected and so painful that he could barely stand to walk. Sandy and I soaked his foot in warm water so we could clean it, and really see the extent of the infection. Using tweezers, scissors and a syringe, we cut, pulled and sprayed away dead tissue. He had a very large open ulcer on the top of his foot and a sinus tract (tunnel) running between his large and second toe to another open area on the ball of his foot. The tissue around the open ulcer was reddened and swollen. After thoroughly cleaning his foot, we packed the open ulcers with charcoal paste and gauze. We also applied charcoal paste and gauze to the surrounding inflamed tissue. We bound his foot with a gauze wrap and put a plastic bag over it to keep out the dirt. We gave him an injection of antibiotics and instructed him to return the next day.

"In my nursing experience, open ulcers are extremely slow to heal. Also, foot wounds are slower to heal than on other parts of the body. The general population there, is typically unreliable when it comes to returning faithfully for treatment until a problem is adequately resolved. So, I did not have high hopes for a good outcome and I feared that this man might lose his foot in the long run. When the man returned the next day for follow-up, I was astounded at the degree of healing that had taken place in only twenty-four hours. The tissue surrounding the open sinuses was far less inflamed, and after a short soak and rinse, the inside of the open ulcers appeared pink and clean. We again packed the ulcers and covered the areas with charcoal paste and gauze as before.

"Clyde and I were only there long enough for me to be involved with this man's care for three days. However, I consider the degree of healing that occurred in just forty-eight hours to be remarkable and unlike anything I have seen in my experience as a nurse."

Whether you are in America, Europe, Asia or Africa, there is little comfort when it comes to being bitten by a poisonous snake, unless … you happen to have caring help who know what to do, including giving charcoal.

On Hand

These stories of pain and suffering represent just a small collection of experiences from "your neighbor next door" as well as professional health workers. But the one thing they all have in common is the outcome. They all demonstrated the efficacy of charcoal as a simple and powerful healing agent.

Once again these experiences demonstrate that the degree of recovery is directly related to the promptness of applying charcoal. If it is available, you should seek professional medical help. However, first use your new emergency anti-poison—charcoal—before you race off.

"Knowledge is power". But it does no good to know about charcoal if you do not have it on hand. For those living, working or traveling in areas that are distant from acute emergency care centers, wisdom would dictate that they should take more responsibility for their own care. Be sure to keep a supply of activated charcoal on hand. If the emergency providers are unfamiliar with charcoal, or if it is not recognized as a primary or secondary line of treatment, then it does not matter if you are rushed in a hi-tech ambulance to a hi-tech emergency unit, or if you limp into a remote bush clinic.

It's like a fire extinguisher. You pay a good price for one that is big enough to do the job, even though you hope you will never have to use it. Charcoal is big enough for the job. You just have to have it available, and use enough of it. In an emergency, I also believe it never hurts to call on the Divine Chemist who formulated it. CR

---

[1] Thrash, Agatha, MD, & Calvin, MD, *Rx: Charcoal*, New Lifestyle Books, p. 70-71, 1988
[2] *Ibid.*, p. 70
[3] *Ibid.*, p. 65, 66
[4] *Ibid.*, p. 68
[5] Boquet, A, Adsorption of cobra venom and diphtheria toxin by carbon, *Comp. Rend.* 187:959, 1928

Chapter 12 Believe It or Not

While charcoal has a well-documented history of its gradual climb to world acceptance, and while it is best known for the benefits to both the industrial world and the field of medicine, charcoal also has a positive reputation for its versatility. This chapter looks at a variety of diseases and infirmities, in no particular order, and how charcoal has proven to be helpful in their treatment.

Tetanus and Diphtheria

I would like to begin with a story that was related to me many years ago. I recorded the story briefly so the details are sketchy. A four-year-old girl was brought into a hospital in Mexico and diagnosed as having tetanus. She was unconscious, and suffering convulsions. The doctor had done all he could for the child but with no effect. One of the nurses who was on duty, asked if she could try a charcoal treatment on the girl. The doctor said there was no help for the girl, but since the charcoal could do no harm, the nurse could try.

The nurse placed the little girl in a bathtub of warm water with a large amount of charcoal powder mixed in. Within half an hour the convulsions had subsided, and the girl had regained consciousness. But shortly after, the seizures began again. When more hot water was added the convulsions ceased. The girl did recover with no complications. Was this just some curious coincidence?

In his book *Activated Charcoal in Medical Applications*, David Cooney reported two separate studies that demonstrated activated charcoal's capability of diminishing, if not neutralizing the effects of the poisonous properties of the tetanus toxin, as well as the toxins of tuberculosis, and diphtheria.[1]

Cancer

The anemia associated with cancer is produced by the cancer's toxic waste products which destroy red blood cells. These toxins retard the work of the bone marrow, but they may be adsorbed by charcoal when taken orally.[2] Certain cancers in dogs have been shown to go into complete remission with the application of charcoal poultices.[3] Then there are those chemicals known to cause cancer, such as benzopyrene and methylcholanthrene, which are also effectively adsorbed, thus preventing their poisonous effects.[4]

Vomiting

Virginia lives in South Carolina and shared one of her many experiences:

> "Some years ago when a friend missed church for several weeks, I called to see how she was doing. Her husband John answered the phone, but said his wife could not come to the phone. I could hear someone vomiting in the background, so I asked if Betty was sick? He said yes, that she had been sick and couldn't see anyone because she had been vomiting continuously. They had been to their family doctor and had then been referred to another doctor. But nothing they had suggested had helped. She was not able to keep anything down.
>
> "I asked if I could come over. She agreed I could but would not be able to visit with me. When I went over, I took some charcoal powder with me. But when I arrived, Betty was in the bathroom throwing up again. She was exhausted and groaned that she had nothing left, energy or otherwise.
>
> "I mixed up some charcoal slurry water, and gave her some to drink. We then sat and had a prayer and waited. After an hour she had still not vomited again, so I decided to leave. Before going, I instructed them that if she did vomit, she was to immediately take more charcoal. But she did not vomit again, and she quickly recovered her strength!!"

Cases of nausea and vomiting, of whatever cause, will almost invariably respond to charcoal taken orally. If the patient vomits after an initial dose of charcoal, a second or third dose will typically settle the stomach.

Foot Infections

For years Nicole has been dispensing charcoal to her family for all manner of things. Not long ago her fifteen-year-old son Andrew suddenly developed a badly infected big toe:

> "I do not know what caused it, whether it was an ingrown toenail or not. It was inflamed, very painful, and about twice the normal size. I mixed up some charcoal with ground flax seed and a little water to make a paste. Then I smeared it all over the toe. The swelling seemed to go down after a few hours. I wrapped the toe with gauze, and he slept with that on overnight. By morning it was completely normal. It was amazing!
>
> "About the same time, my girlfriend's daughter had a very similar condition. She took her daughter to the doctor, and he put her on antibiotics for a week."

As you can see, charcoal's anti-bacterial, anti-septic nature can be just as potent, if not more so, than conventional drugs.

Infected Wounds

Towards the end of my stay in Nepal, Janie, an emergency nurse from a burn unit in California, joined our team. She had seen her share of sad cases. However, sad cases take on a different meaning in a rural setting where there is poor sanitation, and when you are hours from the nearest road and then several more hours to the nearest hospital. Many of the ones who came to the clinic came to it as a last resort. Many were too poor to make the trip to the hospital. Some lived days away from us, and some were too fearful to travel any distance from their villages.

One afternoon a man arrived with his hand bound in rags. Janie raised her eyebrows when the bandages were removed. The man had been in India where he had been caught stealing. His fine was to lose one of his fingers. At some point the wounded hand became infected, and he ended up losing another finger. At this point his hand looked and smelled as if gangrene had set in.

We understood from our translator that the man lived some distance from us, and would not go to the hospital. Realizing it was unlikely the man would return for further treatment, Janie asked if I could suggest some home treatment. All I could think of was charcoal. I was busy with other work, but when I later passed through the clinic Janie had him soaking his hand in a mixture of water and charcoal powder. He was instructed on how to bathe and treat his hand, and then sent on his long walk home. He was told to come back if there were any complications. The man did not return. If he had passed by us to go to the hospital, word likely would have reached us. We were left to hope that the simple treatment had been successful. If only this man, or someone close to him, had known that the medicine he needed was right at hand, and that he didn't have to steal to pay for it, maybe he would not have lost his second finger.

Erna, who we met earlier in the book, is a second-generation charcoal user. She tells this experience. On another occasion:

> "I cut my hand badly and needed stitches. Several days later I still had a fair bit of pain in the wound and I noticed that it was a little swollen and red. I put a charcoal paste on it and within half an hour I was pain free. I left the charcoal on over night and by morning the swelling and redness were gone. I also decided to try Vitamin E oil to prevent scarring, which worked very nicely. I would rub the Vitamin E oil into the wound several times each day. The scar is very fine, even though the wound separated after the stitches were removed. I was expecting a large reminder of the incident."

Pastor Vic is an evangelist that believes preaching and healing are the true credentials of a gospel worker. He regularly travels to Ukraine to conduct programs. While there, he often sees people afflicted with varicose veins. If these become bruised, they frequently lead to ulcers. As a treatment, he regularly suggests charcoal poultices along with simple water baths, and just as regularly he sees improvement.

"For an upset stomach, from whatever cause, I always recommend one teaspoon of charcoal in a cup of warm water." Obviously this confidence with charcoal goes back a long way. Oh yes, Erna is Pastor Vic's daughter. A coincidence?

Vicki and her husband Kevin work in Ireland. Vicki sent on this e-mail:

> "As for stories on charcoal, I have a personal one. Last year while we were in the U.S. on furlough, I had a procedure done by laparoscopy. I had a small incision at the navel. After returning to Ireland, we went on to a retreat in Albania, and while there I did some swimming. Soon after, the wound started looking puffy and red and it was quite sore. Each time after leaving the water I'd clean it but it kept getting worse. After we returned home, almost every night for a week, I made a charcoal poultice, and placed it over the infection. I made a paste with warm water and activated charcoal powder, put it on a square of paper towel, covered it with plastic wrap and put it on the spot. Then I even wrapped an ace bandage around my waist to hold it in place. It seemed to be better each morning and eventually healed up fine. The warmth seemed soothing as well. I was thankful I did not have to use antibiotics."

Big wounds or little wounds, charcoal is ready to stop most infections in their tracks.

Pain

When I interviewed Dr. Wynn over the phone, I asked him what he sees charcoal used for most often:

> "Here at Uchee Pines we use it for many different problems. However, I would say mostly for pain, inflammation and pain of joints, or around the trunk area for bowel complaints. It is most often applied overnight with a flaxseed poultice. The majority of people find relief from whatever pain they are experiencing. But I can't explain how it works. I understand how it might work internally by adsorbing toxins, but exactly how it can relieve deep tissue pain or draw out toxins when used externally is still a mystery to me. I know flaxseed has its own anti-inflammatory properties, but that still doesn't explain such dramatic results."

Pain is a condition all of its own, and when other common forms of treatment fail to bring relief, charcoal can be tried without any worry of doing harm. Mention has already been made of treating the pain of gout (one form of arthritis), but a trial of charcoal poultices should also be made for the pain associated with sore throat, earache, toothache, irritated eyes, sprains, inflammations, and bruises. Relief within the half hour is not unusual. The poultice may be left on for several hours or overnight.

Any area that is red, painful, swollen, and hot often responds to charcoal. The pain produced by cancer, whether in the bone, the abdomen or elsewhere, may often be controlled with a charcoal poultice. If an injury is better suited to a charcoal bath, mix half a cup of charcoal into two gallons of warm water and soak for thirty to sixty minutes.

Tic doloreaux is a very painful habit spasm of the face. But in one case, when we placed a poultice over the affected area of one sufferer, the pain was relieved and the woman had her first full night's sleep in a long time.

Charcoal, it does no harm, it is inexpensive, simple, accessible, and it works!

Pregnancy

Now retired but an active public health speaker, Doctor Calvin, NMD, for many years directed a 22-bed hospital in Montemorelos, Mexico. While he regularly used and recommended charcoal, he explained that his favorite formula was to mix it with an equal amount of clay in poultices. But one experience of using just plain old charcoal still causes him to chuckle:

> "I was coming back through Guatemala on my way to Texas, and who do I see but Linda [the daughter-in-law of another doctor friend] hitch-hiking along the Pan American highway. Of course I stopped, and picked her up. She was very pregnant! She was expecting their second child and, while her husband stayed with their first child, she was headed back to see her family in California.
>
> "She was experiencing severe diarrhea—perhaps a complication of the pregnancy or from dysentery or both. I was having to stop every five minutes to let her have another 'emergency'. Finally I told her, 'We are never going to make it to Texas at this rate.' So I stopped in at some drugstores looking to buy some charcoal, but no one had any. As we were driving on, I noticed a thatched house about two hundred feet back from the road with smoke curling up through the roof. I knew they must have been cooking. So, I stopped and ran down to the house and asked the woman if I could look through her firebox for some coals. She said that it was okay [Well, what would you have said?]. I was able to find a few large pieces. As I blew off the ashes, I asked her what kind of wood she was burning. It turned out to be cedar. I thanked her and took it back up to the truck. I told Linda, 'Here chew on this.' For the next ten miles she did just that, and the diarrhea cleared up. She had no more 'emergencies' all the way back to Texas, where I put her on a bus to California."

War

While we may individually experience a biological war going on in our own bodies, charcoal serves on a much larger battlefield. Both the French and Germans reported the benefits of charcoal in the prevention and treatment of diarrhea during the First World War. But it was the introduction, from 1914 to 1918, of poisonous gases onto the battlefields that gave a great impetus to the development of charcoal's suitability for use in military respirators. What was needed was a material with adequate adsorptive powers as well as low resistance to airflow through the respirator canister. This was achieved by activating wood chips with zinc chloride. These were the first manufactured charcoals with specific adsorptive and physical properties. Coconut shell was the final choice for a starting point.[5]

Here in the 21st century, humanity is still threatened with the barbarism of war. There are renewed threats of chemical and biological warfare. While new, more aggressive organisms and chemicals have been engineered, their natures still make them inherently susceptible to adsorption by charcoal. While the threat of anthrax may be new to the reader, it was not new to Hippocrates or Pliny two thousand years ago. For all the fear that surrounds anthrax, Hippocrates and Pliny both recorded the ability of charcoal to effectively render harmless this extremely infectious bacteria.

Besides defusing biological agents such as anthrax, charcoal has been found to be highly effective in neutralizing mustard gas. Other simple neutralizing agents include mild acids (5% Clorox) and bases.[6]

Apart from chemical warfare or germ warfare, it is worth mentioning again that it is a common practice, in laboratories that perform studies into the transmissibility of different viruses, not to use charcoal-impregnated swabs. More often than not, the charcoal inactivates the infectious nature of viruses. Charcoal's antiviral potential demonstrated in research labs, should translate into one more prevention strategy for personal health.

Air-Borne Chemicals

The following true story, with a little modification, could fit a war scenario. The original letter was addressed to Enos Yoder:

> "I am a certified pipe welder who lives in Columbus, Ohio. In 1994, 1 was working in a refinery in Lima, Ohio. I was working second shift, from six p.m. to six a.m., seven days a week. On the night of October 14, while I was on my way to work, the news report said rain was expected after midnight. About 1 a.m. a fine mist of rain started to fall. About fifteen minutes later a safety man at the refinery came running up. He told the eighteen men in my crew to report to the designated safe tent and wait for further instructions.
>
> "At five a.m., while we were still sitting in the tent, a bus pulled up. The supervisors came in and said that we were all going to the hospital for blood tests. The tests

would be sent to the company and paid for in full by the contractor. What we thought was rain, was actually benzene spraying out of a tower 175 feet in the air. A half mile away, a refinery employee had accidentally turned the wrong valve and exposed us to one of the most carcinogenic substances in the United States. Benzene affects the eyesight, nervous system, kidneys, and liver.

"When we arrived at the hospital we were all given blood tests. These tests revealed that all eighteen of us had dangerous levels of benzene in our blood systems. The doctors at the hospital explained that it would take up to a year for the benzene to leave our systems, and the damage it would do would vary from blindness to liver cancer to kidney failure, and highly possible to death. There was no known treatment to remove the benzene from the blood.

"That morning there were eighteen very scared men, who had come from all over the United Stated to earn a living for their families. Their lives were now being given an expiration date of five to seven years. The next day we were all laid off. The company said they were sorry for what had happened and wished us the best of luck.

"Because of my fulfilling relationship with Christ, that day I prayed. Immediately, the name of a man I had visited a year earlier came to my mind. When I got home I called Enos. He said to just come on up. I did, and he gave me a bag of charcoal capsules. He said to take three capsules, three times daily, and not to skip any capsules.

"After taking the charcoal capsules, I made an appointment with my family doctor. I told him about the benzene and asked for another blood test. When the test came back a week later, it showed that I had no benzene in my system. The doctor said that my case had caused him to spend a lot of time reading about benzene and, according to all he had read, there was no way the body could rid itself of benzene in less than twelve months. He said that the lab knew this also, and they would give me another test for free. Those results also came back, NO BENZENE in my body!!!

"About six months later, the contractor refused to pay for any testing done to me. This led to a court action because federal law, workman's compensation, and O.C.C.A. regulations said the contractor was fully responsible. Approximately three years later I went to court.

"Upon arriving, I met a man who was scheduled to appear in court at the same time as I was. After introductions, it turned out that he was with a law firm representing seventeen men who had filed a class action suit due to benzene poisoning in Lima, Ohio in 1994. The court did not know that I was not part of that lawsuit, and had scheduled me at the same time in the same courtroom. The attorney explained to me that two of his clients suffered from blindness, four were on kidney dialysis, and all were being treated by a neurological psychiatrist. The lawyer was amazed that my benzene level had been more than twice that of any of his clients, yet I had no health problems whatsoever. He said that it was nothing short of a miracle for me.

"To that I first said, 'Thank you Jesus for watching over me,' and second, 'Thank you Enos Yoder for being there, and caring the way you do.'"

Eyes

"Pink eye", or "sore eyes" is an acute form of bacterial conjunctivitis—an inflammation of the mucous membrane which lines the inner surface of the eyelid and continues over the surface of the eyeball. This condition is highly contagious and often occurs in groups of schoolchildren. Conjunctivitis is probably the most common eye disease in the Western Hemisphere. Symptoms include redness, swelling, tearing and discomfort. There may be a discharge from the eye. Generally a thin, watery discharge suggests that the conjunctivitis is of viral origin, a white, stringy discharge suggests allergic origin and a discharge containing pus suggests conjunctivitis of a bacterial origin.

The conjunctiva has relatively few pain fibers, so a sensation of discomfort, burning, or scratchiness is common. Itching and light sensitivity may also be present, especially in allergic conjunctivitis. In their book *Natural Remedies*, the Doctors Thrash outline the following treatment:

1. Charcoal poultices should be applied overnight. Mix powdered charcoal with water sufficient to make a thick paste, and spread it over a piece of flannel or muslin larger than the inflamed area of the eye. Place this over the eye. Cover with a piece of plastic or similar material and hold in place with an ace bandage wrapped lightly around the head. The bandage should not be so tight that it puts pressure on the eyeballs, but it must be snug enough to hold the compress in place overnight. Remove it in the morning, and dispose of the compress in a manner to avoid spreading the infection.
2. Charcoal slurry water eye drops may be used during the day. To make the drops boil one cup of water with one-fourth teaspoon salt and one teaspoon powdered charcoal. When cool, strain through several layers of cheesecloth. Using a dropper, put four or five drops of the clear fluid in the affected eye every two hours.[7]

When Jeff and Bobbi moved to Papua New Guinea they settled into a village of about one thousand people. Pink eye was so prevalent that each family had one or two members who suffered with it. It is very contagious. Over the phone, Bobbi told me that she developed pink eye not long after arriving there:

> "People were so friendly and eager to shake our hands and hug us ... Shortly after we got set up, Ulato, a mother of two, came to us with a quilt over her face. Her eyes were so sensitive to light she couldn't stand to have them uncovered. We gave her the charcoal water to take, and by the fourth day her eyes were clear. She had been in so much pain.
>
> "Gege was a young mother of three boys. The middle son had pink eye so bad that his eye was swollen shut. She told us the charcoal was not working. It seems that she was just dabbing a cloth in the charcoal water then dabbing the corner of the eyes. But after we explained to her just to pour the mixture and to let it wash over the eye, her son was cured within days. Most of the cases had a discharge from the eyes and the white of the eyes was an angry pink.
>
> "We tried the standard antibiotics but found the charcoal to be ten times as affective. In fact, when we left, the whole village was virtually free of pink eye, and the people only wanted charcoal to treat it."

Bobbi used formula 2 (page 133). But, because of the circumstances, she gave charcoal to the people in a jar, and simply instructed them to add water, then shake it vigorously, and let it settle out until it was "see through". Then it was ready to use as an eyewash.

Ears

Novel serves as mother, department head, counselor, and singer. She is also a "private" nurse. "My first experience with the healing properties of charcoal happened when my infant daughter had an ear infection. Initially I took her to the doctor for the usual treatment. He put her on a penicillin-family derivative and her ear infection cleared up.

However, after a few weeks the infection returned. When I mentioned this to a girlfriend, she suggested to me that putting a watery paste made from activated charcoal, into my daughter's ear might very well clear up the infection without resorting to several doses of antibiotics. My girlfriend came over to my home and demonstrated how to make the thin paste. Using an eyedropper, we carefully dribbled it into my daughter's sore ear. The ear infection soon cleared up and never returned. I took my daughter back to the doctor for a final check to be sure everything was clear inside. He assured me that there was no longer any infection." Novel told the doctor about the treatment she had done and he was supportive.

Pleurisy

Novel continued:

"My next experience with charcoal was when my husband developed pleurisy. He was experiencing some discomfort and chest pain so we had to take him to emergency at the local hospital. After ruling out a heart attack, the doctors x-rayed and said that he had pleurisy and immediately started him on an antibiotic treatment. We brought him home, but the antibiotics were slow to get under the infection. It was then that I remembered about charcoal, and decided to do a poultice treatment of hot charcoal and flax seed [as a thickening agent] on his chest and back. After several treatments over a couple of hours, my husband began to have relief and felt more comfortable. The charcoal drew the infection from him and the pleurisy was cleared up. It has never returned."

Sore Throat

Novel also has a son:

"Until recently, my use of charcoal was limited to external treatments. However, when my twenty-one year old son developed a severe sore throat and flu this past Christmas season, another friend advised me to have him slowly swallow a thick charcoal paste mixed with water. Normally we do not worry too much about sore throat or cold symptoms and generally just 'weather the storm.' However, the fact that our son was having difficulty breathing was a serious matter. Two years before he had had a severe accident that paralyzed his larynx, and left him with limited breathing capabilities. Therefore, it was extremely important that he quickly be made as comfortable as possible. The over the counter throat lozenges and syrups were not helping. So, when my friend suggested getting my son to eat a charcoal paste, I immediately went to the cupboard and made some. Within seconds of having swallowed the paste, my twenty-one-year-old son called out that he was breathing comfortably, and that his cough was relieved. He continued taking the charcoal paste periodically for the remainder of the day, allowing it to slowly trickle down his throat.

"I always have activated charcoal on hand and often advise my friends of its wonderful healing properties. Now I cannot only vouch for its external benefits, but also for its internal value."

Once you have experimented with this harmless remedy, it won't be long before you too will be recommending it to others.

Diabetes

Diabetes mellitus involves many factors. If one has diabetes, one's lifestyle will either be its worst poison or its best medicine. But as a simple adjunct to any treatment program, charcoal has been shown, in some individuals, to reduce glucose levels in the bowel resulting in a reduced need for medication. It has also been shown to reduce the signs of blood vessel and nerve damage—a major complication of diabetes.[8] While obviously not a cure for diabetes, some may want to consider charcoal as an aid in managing their cases.

Allergies

Allergies are both varied and perplexing, and they should not be treated any more casually than any other ailment. But, while trying to ferret out the cause of some debilitating sensitivity, it may be worth experimenting by giving some activated charcoal orally. It may do nothing to bind the allergens themselves, as with certain viruses and bacteria, but it may neutralize other secondhand toxins produced by the allergens.

Aflatoxin B_1 and T-2 toxin are both fungal toxins that kill both humans and animals. Once again charcoal is a powerful antidote for these and other fungal toxins.[9]

Peanuts

In a recent study by Peter Vadas, MD, PhD, University of Toronto, and colleagues, it was found that administering activated charcoal, in addition to epinephrine, soon after the accidental ingestion of peanuts, may help to reduce the severity and progression of life-threatening anaphylaxis. The Vadas study showed that activated charcoal will bind within sixty seconds to the major allergens in peanuts, preventing remaining peanut proteins from activating an allergic response. The conventional treatment of peanut-induced anaphylaxis is only directed at treating the symptoms by administering epinephrine, antihistamine, a steroid, and stabilizing breathing and circulatory functions. To date there have been no therapies directed toward slowing or preventing further absorption of peanut protein from the gastrointestinal tract after an accidental ingestion. This study demonstrated in a remarkable way the ability of activated charcoal to form a complex with peanut protein, thereby diffusing an allergic reaction. A ratio of 200 mg of activated charcoal to 1 mg peanut protein was required to completely remove the peanut protein from solution in lab experiments. Activated charcoal was also able to bind with peanut protein when peanuts were mixed with other foods such as ice cream and chocolate. This approach to preventing fatal reactions may also prove useful with other allergenic foods.[10]

After this amazing study was published there was a swift response from one interest group warning that charcoal is no substitute for antihistamines or epinephrine in the treatment plan for severe allergic reactions to peanuts and other foods. They stated charcoal slurry "tastes awful", may interfere with medication, children may not accept it, and it will not stop a reaction in progress. The detractors seemed to overlook the fact that the study used charcoal in conjunction with epinephrine, and it did show that charcoal slowed the progression of the reaction. The other concerns raised are reminiscent of the uninformed and empty fears promoted twenty years ago by those clinging to the efficacy of ipecac in the management of accidental poisoning in the home. Today, activated charcoal is the treatment of choice for accidental poisoning in the home. One can only hope this new information will not be summarily ignored for the next twenty-some years while victims continue to suffer.

When dealing with severe allergic reactions, as with any poisoning, time is critical. How much wiser it would be to take charcoal after accidentally ingesting some known allergenic food, while proceeding on to the nearest emergency facility. Changes within medical tradition grind very slowly. In the meantime, those who know from long experience what they are talking about, have in charcoal a remedy they know will do no harm.

Julie tells the case of one gentleman who attended a lifestyle program at Wildwood Hospital. Jim had a severe allergy to dairy products, and ate a cookie that triggered a reaction. Typically, within twenty minutes, he would become very bloated, and have a terrible migraine headache. Before he came to the Lifestyle Center connected to the hospital, he called and was told to take one tablespoon of activated charcoal in water. He followed directions and never did experience the headache, even though he did have some bloating.

It is reassuring to know, when one is unsure about what remedy will work and what will not, that there is no worry of compounding problems with charcoal as there is with most drugs.

Tom has worked for over the past ten years with people who were seriously ill. He has seen the benefits to very sick patients who used charcoal slurry drinks in large amounts (up to twelve glasses per day). He reports they would almost always get well again. "Since activated charcoal powder acts like a magnetic sponge to quickly eliminate most toxins from the body while it's detoxifying itself, the body can better handle the much lower amounts of toxins which remain. Myself? I've had several cases of severe food poisoning and drank plenty of charcoal slurry, which, of course, alleviated my serious health condition each and every time!"

Toes

Here is a hot-off-the-press down-to-earth story from my sister. Deanna has worked and traveled extensively as an educator. Teaching in First Nations communities throughout western and northern Canada, she is familiar with native remedies and their stories. But charcoal does not seem to be a prominent one in their history:

"I have for most of my adult life, suffered with ingrown toenails. Many friends, doctors and podiatrists have given me sure-fire cures. None of them have worked. So, regularly, I schedule a quiet, but increasingly with age, uncomfortable session by myself, digging with sharp cuticle scissors at my poor misshapen toe, trying to find and extract the offending bit of rogue toenail. Mostly I am successful, but lately I find the chore terribly irritating and strenuous. So, I too often just quit before the operation is successful. Invariably, this leads to an exceedingly sore toe—and frequently to infection.

"This happened last month when, within hours of my "poking around", the angry red streaks had moved from around my toe up onto the top of my foot. I hate taking antibiotics, and I know that if I had to consult a doctor, that that would be the unquestionable outcome. So, being in the middle of editing your book on charcoal and its amazing properties, I decided to try that.

"Our drugstore had no charcoal in stock, and in fact our young, arrogant pharmacist was apparently horrified that I would even consider trying to "cure" myself. I then drove over to our nearest small city, and there in the health food store, I found powdered charcoal. However, it wasn't until bedtime that I was able to actually mix up my paste, and wrap my toe. It all looked quite "hokey", with bits of charcoal dribbling from my wrapped and re-wrapped, and bandaged toe. But off to bed I went trying hard to keep my by-now throbbing foot, still and quiet.

"I woke in the morning with a wonderfully uninfected toe. However, I also had charcoal from one side of my bed to the other. The heavy paste had apparently dried out as it fell off my foot, and had rolled with me as I tossed from side to side in my bed.

"The result? My toe looks happier than it has in years. It is not at all red, and somehow the ingrown toenail seems to have "loosened up". However, I do, I think, have permanently gray sheets. Next time I will take greater care with my wrapping and bandaging."

I guess my warnings about being careful when handling charcoal were not as forceful as they should have been. So take some sisterly advice and remember, prevention is better than cure—for bedding and clothes as well as toes.

In Review

As we review what we have learned this far, we know that: charcoal was used for preserving Egyptian mummies, fence posts, and water around 3500 B.C.; charcoal was used for the relief of numerous ailments recorded by Hippocrates and Pliny from 400 B.C. to A.D. 50; charcoal was recorded in hundreds of treatments for a wide range of diseases by Claudius Galen, A.D. 157; charcoal was used for purification in the food industry beginning in the late 1700s; charcoal's use came of age in the 20th Century in science, in industry and in the health industry.

From space stations to fuel cells, from hospitals to homes, from tetanus to toes, charcoal's reputation is built on an unshakeable foundation. There is no man made product throughout man's history that can hope to catch up to the successes of this little giant. With the astronomical discovery of C_{60} in 1985, charcoal research has been launched into a stellar orbit. The ability of these concentrated charcoal components to trap and retain gases, and other chemicals, at such enormous ratios, as one researcher said, "can only be described as extraterrestrial in origin"[11] Charcoal is making waves around the world.

Is there more? CR

---

[1] Cooney, David O, *Activated Charcoal*, TEACH Services Inc., New York, p. 53, 1999
[2] *Journal of the American Medical Association*, 237 (17): 1840, April 25, 1976
[3] *Journal of the American Medical Association*, 54:331, December 7, 1910
[4] *AMA Archives of Industrial Health*, 18:511-520, December, 1958
[5] Carbon Materials Research Group, Center for Applied Energy Research, University of Kentucky, Historical Production and Use of Carbon Materials, December 5, 2003
[6] Chemical Warfare Agents: Their Past and Continuing Threat and Evolving Therapies, http://www.medscape.com/viewarticle/461822; Common Chemical Agent Threats, *Neurosurgical Focus*, http://www.medscape.com/viewarticle/431312
[7] Austin, Phylis, Thrash, Agatha, MD, & Calvin, MD, *Natural Remedies*, p. 19, 1983
[8] Thrash, *Rx: Charcoal*, p.42, 1988
[9] Cooney, David, O, *Activated Charcoal in Medical Applications*, Marcel Dekker Inc., p.478-480, 1995
[10] Vadas P, MD, Perelman B, "Activated charcoal forms non-IgE binding complexes with peanut proteins." *Journal of Allergy & Clinical Immunology* 112 (1):175-9, July 2003
[11] Luann Becker*, Robert J. Poreda+, and Ted E. Bunch#, *School of Ocean and Earth Science and Technology, University of Hawaii; +Department of Earth and Environmental Sciences, University of Rochester; and #Space Science Division, National Aeronautics and Space Administration Ames Research Center, Fullerenes: An extraterrestrial carbon carrier phase for noble gases, *Proceedings of the National Academy of Sciences US*, vol. 97, no. 7, pp. 2979-2983, March 28, 2000

Chapter 13 From Head to Toe

The following stories demonstrate once again, how adaptable charcoal can be, no matter whether it is in its more crude form or when it is activated. I have to wonder, after listening to so many different stories of its applications, if the only areas charcoal offers no benefit to at all, are those we have supposed that it can't work for. From the crown of our heads to the soles of our feet, we are daily assaulted with injuries and maladies. But, whether at home, on the job, or far away, if we are not far from charcoal, we are never far from help.

Injuries

As an accountant, Larry's work usually confines him to indoors, but he enjoys being outdoors—when its safe. Larry tells of one of his several outdoor collisions:

> "I received a head injury from a steel wedge that I was using to split wood. It popped out of the wood and hit me on my head, giving me a bad gash. Afterwards, because my eyes would not focus properly, the person checking me out was very concerned that I may have some problems. So, I was rushed to the hospital in an ambulance. Before the ambulance arrived Roi put pine pitch in the open wound, and that helped tremendously.
>
> "After my release from emergency, I returned home to recover. I still had some not too serious problems, but the throbbing pain in my head intensified. Since I needed to be well before flying off to Hawaii in a week's time, here is what I did. I put a large charcoal poultice on my head, covered that with plastic wrap and then put a cap on to hold the poultice in place. As long as I wore this poultice I didn't have any major headaches or problems. But, when I experimented by taking the poultice off my head, I started to have problems again. So, I kept the poultice on my head for the whole week, and when I took it off just before we flew out, I was better."

Roi is the one who ends up nursing Larry after his adventures:

> "I also used a charcoal poultice on Larry when he had his motorcycle accident and hurt his shoulder. I put a large charcoal poultice over the shoulder area overnight and it really helped with the bruising. It seemed

to miss the black and blue stage and go directly to the healing green color. Anyway, I thought that it helped very much with the bruising."

Sherman works in a hospital office. When he was eighteen, he had a serious injury to his lower leg. The people caring for him made a charcoal and flax seed poultice and applied it over the injured area. They also applied mild hot and cold towels. By faithfully applying these alternate treatments, Sherman's recovery was quick and complete.

Cheryl and her husband work at Black Hills Health and Education Center in South Dakota. I was able to catch her before she headed off to class. Cheryl shared her story over the phone:

"Even though I have one leg paralyzed from polio, that left me a paraplegic, I have lived a very full life. I am a concert violinist and have taught for thirty years. My husband is a retired dentist but for years we worked and traveled overseas. I have two sons, and have worked my whole adult life. Now I work here at Black Hills in the lifestyle center, and my lecture on charcoal is my specialty.

"I am now fifty-seven, but some time ago I began to develop cellulitis on the shin of my paralyzed leg. The skin would become fiery red and very sensitive. Because sepsis can set in and attack the whole body, cellulitis can become a deadly infection if left to run its course. In fact, on one occasion it did land me in the hospital. It had been bothering me for several years, but the doctors could only recommend stronger and stronger antibiotics. However, the antibiotics seemed to work less and less effectively, and I was afraid they would become ineffective if I ever really needed them.

"At the time, my husband was practicing at a California hospital, and so I finally went to the orthopedic surgeon and asked if there was anything he could recommend. He directed me to a vascular surgeon. On my first visit to this specialist, based on his brief examination, he said, "I don't know why you don't just get it cut off!" referring to my leg.

"So much for that. I continued taking antibiotics, but I also began to ask around for some alternatives. Three years ago, when I was visiting near Wildwood Hospital in Georgia, I was directed to a doctor who worked there. I asked her how I could keep the cellulitis from flaring up again.

"'It is so simple!' said Dr. Robbie, MD, 'All you need is two buckets, two thermometers and some hot and cold water.' She then showed me how to give myself hot and cold soaks to the leg, and how to apply a charcoal poultice over the affected area overnight. Within a couple of days I had

> my healthy pink color back again. For the past three years I have not been bothered with the cellulites. I have found that at the first sensation of it coming on, if I immediately begin to give myself the hot/cold treatments (seven changes), with the charcoal poultice overnight, the cellulitis never develops."

Here is the recipe for charcoal paste that has worked best for Cheryl:

- One part psyllium seed husk
- Three parts activated charcoal powder
- Just enough hot water to make it into a dough consistency.

This can be rolled out on a piece of saran wrap about an eighth of an inch thick, then sealed well, and placed in the fridge. For several days use, she mixes a quarter cup psyllium with three-quarters cup charcoal. If she is scheduled to travel, she takes some of this with her just in case. Incidentally, charcoal is known to work well for cellulites of the face, eyelids and ears, and we can only assume elsewhere just as effectively. The rapid relief from pain is hard to appreciate unless you have experienced it for yourself.

Teeth

Here is an assortment of tooth stories that are better than fairy tales.

While in Nepal, Joel and Joyce cared for a six-year-old girl that had a horribly abscessed tooth. Her cheek was terribly swollen. Joyce sewed up charcoal into little pouches (teabag-size) and put them into the child's mouth. These were changed every half hour:

> "We told her to go and have her tooth pulled. But the charcoal took care of her pain and the infection, so her family felt there was no need to go to a dentist. But in two weeks there she was back again. Only this time the abscess had ulcerated right out to the surface of the jaw. It was a horrible sight. We repeated our treatment using the teabag-size charcoal poultices. Again the abscess improved, but this time she did go and have the tooth removed."

However, dental nightmares aren't exclusive to developing nations or to unsanitary environments.

Helen is an RN. Over the years she has treated herself and her family with charcoal. They have also treated people that they have brought into their home:

> "My first experience with charcoal was when I had my four wisdom teeth removed. The pain was intense. A friend suggested charcoal. I took four

charcoal tablets and positioned one at each spot where a tooth had been removed and held it there with the cheek. The pain left almost immediately. As they dissolved, I would replace them. I continued this routine until the gums had healed. I have since used my experience to help others.

"Cyndy was thirty-seven years old, and scheduled to have all of her teeth removed. Her entire mouth was infected, and the dentist had drained two cups of pus from the gums. The pus was compromising her health, so he scheduled her for an appointment in two weeks. Cyndy was not ready to lose all of her teeth.

"We suggested fever therapy to help raise her immune system, but she was not consistent in following our directions. She ended up coming to stay with us. It was such a severe case that we had to make a very thick charcoal gel and pack her cheeks with it. She had to drink through a straw and, except for meal times, we kept her mouth packed with the charcoal. She accepted this treatment, and after one week the infection was completely gone. There was no need to remove even one of her teeth! The dentist was completely amazed."

Pastor Vic related to me that, during a series of meetings he was conducting in Manitoba, Canada, he was in serious pain with an abscessed tooth. A church member, a nurse, made him a poultice with charcoal and flax seed (half and half) mixed with enough water to make a paste and wrapped in cheesecloth:

"When I went to bed that night, I placed it inside my mouth next to the affected tooth. About three a.m. I got up and took out the poultice. When I woke that next morning the pain was gone, and I have not been troubled with that again."

John chipped a lower rear molar at a time when their income didn't allow for an expensive visit to the dentist:

"Dental insurance from my work did not kick in for two more months, so I decided to wait it out till then and have it capped. But the root ended up dying, and the pain was not going to wait. But I used charcoal to tide me over. I placed a charcoal tablet between the tooth and the cheek and kept one there until I got to see the dentist."

Ruth is an RN and public health lecturer:

"Before leaving for India I broke a tooth. Since there was no time to see a dentist, and because it was not bothering me, I decided to go. But on the

> second day after arriving in India it began to throb. By the next morning my mouth was quite swollen. I made a small charcoal poultice wrapped in a small piece of cloth, moistened it and placed it up against the abscess. I did this for two days. I noticed pus on the poultice when I removed it, and discovered a small drainage hole in my gums. The pain and the swelling disappeared, and I was not troubled with the tooth again until after I returned to the States. Then I had it looked at by a dentist."

Mouth Sores

I mentioned earlier about Bobbi's experience of introducing charcoal in Papua New Guinea. Bobbi told me:

> "The biggest problem was getting them to drink charcoal water. In fact they wouldn't. So we weren't able to help them with gastrointestinal problems. But many had mouth sores and eventually we were able to get them to rinse their mouths with charcoal water. It worked well."

When I was small my mother led me to believe that sores in my mouth were the sure result of speaking some untruth. Those sores pricked my conscience for days. If I had only known about charcoal. Yes, canker sores, those small ulcers in your mouth, will respond nicely to charcoal. Just holding a charcoal tablet up against the sore with your tongue will remove the soreness within minutes. Honest, it's no fib.

Toothpaste?

If charcoal can whiten sugar and remove unwanted colors from a host of products, what can it do with badly stained teeth? As we read earlier, charcoal was prepared commercially as a tooth powder a hundred years ago. For polishing and cleaning, there is nothing superior to charcoal. Just wet your toothbrush and dip it into a small glass jar of charcoal powder. Your toothbrush and your teeth will look black, but it rinses off. Although it is a little messy, it can remove tarter and plaque build-up. After a swish or three you will notice your teeth are definitely brighter. It also helps with bad breath. You will be able to taste the difference, and, I am told, it makes kisses sweeter.

Needle Abscess

While Ruth was in India, she heard this report. A minister from Mizurum gave a charcoal demonstration on Sunday to a church group. One of the participants was a nurse who worked at the public hospital. When she reported for work Monday, she found a man who was suffering a great deal of pain. He had been receiving some intravenous medication when the needle had removed partly from the vein, allowing the medication to drain into the tissue. Whatever the medicine was, it caused the surrounding tissue to die and created a great deal of pain. The nurse suggested to the doctor that they put charcoal on the affected area. The doctor asked her where she had heard of such a thing. She said she had learned about it just the day before at the health lecture. Since it would certainly do no harm, the doctor said she could try it.

Since they also cooked with charcoal in the hospital, she went to the kitchen and gathered some up. This she placed in a pillow, and pulverized it as best she could. She then wet it and wrapped the entire arm with the pillow poultice. She replaced the poultice once before leaving after work. Even by then the man was experiencing relief. By the next morning the redness was gone, the pain relieved, and most of the swelling was gone too.

Geriatric Cases

If you have visited a senior's home no doubt you have seen some of the elderly with ulcers on their feet or legs. Circulation to the extremities gets more sluggish with age, and so, some otherwise innocuous sore can easily turn into an ulcer. Whether the ulcers are treated or not, they sometimes continue to worsen, resulting in amputation. It is so sad especially when there is such a simple solution. The next six stories all begin with serious cases of ulcers in the extremities and end with complete recovery.

While living in Nova Scotia, Canada, Pastor Vic met a woman in her eighty's who had had an accident while getting out of a car. Her leg was scrapped badly and the scrape had turned into an ugly ulcer:

> "By then she had been going to doctors for some time. She had been using salves and creams and bandages but because of her age, the ulcer continued to worsen, growing to almost three inches in diameter and deepening almost to the bone.
>
> "My wife Netty and I came over and began making charcoal poultices and regularly changing them. Soon the infection stopped and new flesh began to fill in. Within a couple of weeks the ulcer had shrunk to only dime size. Eventually it healed over completely.
>
> "At one old folks home, another elderly lady was scheduled for amputation of her foot because of gangrene. Netty began giving her hot and cold treatments and charcoal poultices, and again the foot was saved."

Of course Pastor Vic would be the first to tell you they prayed earnestly for God's added blessing, but he also firmly believes, "faith without works is dead".

This is the second of Pamela's stories:

"For a while we operated a family-type care home. There was one elderly lady who developed a pressure sore on the top of one foot. It would heal over, and then would open up again. We had home health nurses coming on a regular basis to give her the regular treatment. They would soak and debride the wound, and then apply some medication. For one year it never healed. It was about a half-inch square in size.

"It was then we moved out of town to the country, and we began to take a few clients into our home. One of them was this elderly lady. The home health nurse came once a week, and tried to keep the sore from getting worse. He was leaving for a couple of weeks and told me how to treat it. But once he left, I decided to try my own thing. I made some charcoal jelly and placed that directly on the wound. I covered it with a piece of plastic, and pulled a sock on over that.

"The next day when I removed the bandage, the whole surface of the sore had sloughed off and there was nice new pink flesh in its place. There was a hole about the size of a marble with dark reddish skin around it. The second day the lady said, "It's not hurting anymore." The third morning it was healing from the inside out, and it was back to normal by the time the nurse arrived again.

"'What did you do?' he asked. 'Oh, just what you told me,' I answered. 'That didn't cure that. Now tell me what you did,' he insisted. He was so impressed that he took the story back to his office and shared it with the other staff members.

"Now, when I see cases like this in nursing homes, it makes me grieve to know they could be so easily treated. They don't have to suffer. They don't have to get worse, and they don't have to have amputations."

After her experience with a brown recluse spider, Emily was encouraged to use charcoal on someone else's leg. She shared this second remarkable story about her new friend:

"I will call her Sal. A few weeks after we met, she told me about a festering wound on her leg. She did not know how to heal it, and her doctors had prescribed topical antibiotics that had not worked. That very night I went to look at it. The circular wound was one and a half inches wide. There was about a four-inch wide area surrounding it that was inflamed. The wound had necrosis setting in. She is a diabetic and had swollen lower legs where the wound was located. There was a scattering of smaller infected wounds that also needed intervention.

"Since it was an open wound, I protected myself with medical gloves. Then I proceeded to apply alcohol and peroxide to open and clean up the tissue. Because of the loss of circulation, the wound did not even hurt, so I also attempted to remove the top layer of the infection with scrubbing. She felt that a little, so I applied an ice pack just above the wound to help numb the area. I cleaned out all of the wounds in the same manner. Then I applied a charcoal/ground flaxseed poultice. I used aloe vera juice with blended comfrey to wet the mixture. After wrapping her up, I put cotton socks on both her feet to hold the dressings on overnight. This was done for one week with visible improvements each night.

"I knew that establishing circulation was critical. After three nights, the skin surrounding the area peeled off, and new healthy skin started to show through. By the end of the week, the whole area showed new healthy growth. The following week, I taught her to do the treatments herself. By the end of the third week, there was a healthy scab in place. We always ended with prayer and thanksgiving."

Emily's interest in health seems to run in her family of nurses and doctors. But, when her children were born with heart complications, her interest jumped into high gear. In addition to the full-time responsibilities of caring for her family, and home schooling her children, she has also assumed the role of primary healthcare giver. Consequently she became very knowledgeable about the adverse affects of drugs, and has taken more responsibility in learning about and using natural cures. I asked Emily a couple questions about their experience:

"'Have you given charcoal to your children before?'

"'Several times each.'

"'How did they react?'

"'Charcoal has texture, not taste. They like the affects and they just look funny with black mustaches after drinking it. Occasionally I will add liquid chlorophyll and vegetable glycerin to it. That sweetens it and adds green color to it that the kids think is very funny.'

"'Does your husband use it?'

"'Yes. He has started asking me for it anytime his stomach is upset or he has heartburn. He resisted using charcoal the most, but is very happy to drink it now."

As Emily has continued to minister to her family, and also to those who have not been helped by conventional medicine, her faith in natural remedies has grown. And, when it comes to charcoal, she certainly is not alone.

Lucille lives with her family in New Brunswick, Canada. Like many caring mothers, she not only nurses her own family, but often meets those who need that extra bit of personal care. While doing some home nursing for a lady, Lucille noticed that her 84 year-old father had an open ulcer below the knee in the calf of his left leg:

> "It was not healing. I would say it was about loonie-size (Canadian dollar coin) and infected. The swelling was twice normal size. He was being seen by a home care nurse and had been receiving antibiotics. But, the nurse had told her that the sore was not responding to their treatment and that when it turned purple/black to call them. They would have to consider amputation.
>
> "I was reluctant to get involved, but realized it was critical. So, I suggested making a charcoal and flax seed poultice, and then laying the moist side against the skin. Over this they could place some of the blue plastic bed sheet used by the hospital, and wrap it with elastic bandage to hold everything in place. The daughter followed my suggestions and the swelling promptly went down. The ulcer began to turn white, and the circulation improved. I also suggested they bathe the ulcer with garlic water [a mild antiseptic], four cloves of garlic boiled in water. The ulcer was to be washed twice daily. Then two-inch strips of aloe vera plant were to be placed over the open wound. The daughter was able to follow my suggestions. The leg was not amputated, and he was spared a great deal of suffering."

As we have seen, the only real expense that comes with using charcoal instead of drugs is one's time. But the rewards far out reach the sacrifices when one realizes just how much relief of body and mind has been gained. I must add, that one really isn't finished with giving these simple treatments until the patients have been taught how to do the treatments for themselves.

Odors

Foul odors fall into several categories; halitosis or bad breath stemming from poor oral health or indigestion; flatulence or intestinal gas; colostomies or ileostomies; infected wounds; and odors from body casts.

In cases where individuals have lost part of their bowels, gas odors can be problematic. Some take charcoal internally to help control flatulence. Others use stoma bags that are specially designed with charcoal filters to adsorb any odors.

Inflammations can abscess and, along with injuries and postoperative wounds, can become infected. They in turn can produce very unpleasant odors. These odors are the result of destruction of the tissue by bacteria. Just as we have seen how effective charcoal is in neutralizing artificial gases, it is also very effective in controlling wound odors. But it goes even further than just adsorbing the foul smells from wounds. Charcoal not only binds the odors, but it may also stop the very process of decay that causes the odors.

While working in Nepal, Joel encountered a regular stream of infections and abscesses. "One older teen had a deep tropical sore about a half dollar in size. Pus oozed from the ulcer, and it was really foul smelling. I cleaned out the rotten flesh, and just poured dry charcoal powder into the wet wound. By the second day you could see healing had begun. The stench was gone, the wound was clean and new pink tissue had started filling in."

For those of you who have suffered a broken limb that required a hard cast to immobilize it, you are no doubt familiar with the bad odor that develops. Most often the smell is just from dead skin, but it may be from an open draining wound. These odors are not only unpleasant, they are themselves toxic, and they slow the healing process. This requires that the casts be changed often.

To avoid such frequent changes, Dr. Frank Haydon, MD, at Fort Benning, Georgia, developed a simple technique. He took fifteen grams of activated charcoal (about three to four tablespoons) and mixed it with enough water to make a slurry. After the first layer of cast was applied, the charcoal slurry was then poured over the area of expected drainage. The remainder of the plaster was then applied over this wet charcoal. The cast appeared slightly gray, but was accepted well by patients. The unpleasant odor of draining wounds was controlled for much longer, and there were no adverse effects on wound or fracture.[1]

The Doctors Thrash relate this case of an overdose of X-rays:

> "We had a patient who had a large, deep ulcer (twelve inches in diameter) due to an x-ray burn on his back. The burn was from an overdose of x-rays used for treating a skin cancer. The ulcer became infected and foul smelling. His entire house smelled of the ulcer, despite the most fastidious care. We started dressing the ulcer by sprinkling dry charcoal from a saltshaker on all the moist areas before applying gauze. Instantly the odor vanished from the ulcer, and gradually left the house. Although the patient eventually succumbed to the radiation sickness, he and his whole family were grateful for the charcoal."[2]

The foul smelling wounds caused by the bacteria Bacillus pyocyanase can be dissipated in one treatment by the use of charcoal.[3] Healing is also improved, as the charcoal takes up bacteria. One good way to apply charcoal to a foul ulcer is by putting some of the powder in a saltshaker with a few grains of rice. With every bandage change, shake the powdered charcoal on the wound. Eczema has also been successfully treated using charcoal, especially where there is infection and odor.[4]

The prestigious British medical journal, *The Lancet*, reported this exciting study. In varicose leg ulcers and in infected surgical wounds, a single layer of charcoal cloth covered with a porous fabric sleeve dressing gave a noticeable reduction in wound odor in 95% of 39 patients. Wound cleansing was also noted in 80% of the patients. There were no adverse reactions to the material. The dressings did not stick to the wounds and could be removed without difficulty. Because the human skin allows for the transfer of liquids, gasses and even micro-particles through its permeable membrane and pores, it was also shown that warm, moist activated charcoal poultices were actually able to draw bacteria and poisons through the skin and into the poultice.[5]

There is a foul odor that comes with inoperable cancer of the cervix. This too can be speedily remedied with a solution of two tablespoons of charcoal powder to one quart of water given as a douche.[6]

Then there are foot odors. One reputable footwear company is now marketing a patented gel insole that is layered with super-activated charcoal. There are several other companies offering a variety of footwear products with charcoal to combat odor and promote general relaxation.

As for halitosis, charcoal helps to eliminate oral odor, because it cleanses both the mouth and the digestive tract. Since the main cause of bad breath is found in the mouth, swishing some charcoal around in the mouth will promptly neutralize most offensive breath.

In the case of an upset stomach after eating, whether one overate, ate wrong food combinations, ate too fast, ate meals too close together, ate too late at night, or ate food that was too old, you may be benefited by charcoal as a health aid. The foul odors produced by putrefaction in the stomach will be quickly adsorbed by charcoal.

Gas

Suffering from just plain gas? Michael Levitt is the gastroenterologist at the Minneapolis Veterans Affairs Medical Center. Touted as a world authority, he has authored dozens of articles on the subject of flatulence. While it is undisputed that charcoal taken internally dramatically decreases the volume of gas, it does not manage the odor as well as some would want. So he developed a seat cushion to see how much of the remaining offending sulphur (the chief offender) was adsorbed. After a hefty meal of pinto beans and lactulose (a poorly absorbed sugar), "to enhance output", volunteers were dressed in gas-tight Mylar pantaloons to collect what managed to get through the cushion. "I didn't think the activated charcoal layer would be adequate to absorb all of the sulfur," he recounts. "It was only a thin layer. But it worked."[7]

Products

You may laugh, but different companies have developed underwear layered with charcoal. One was initially advertised as the Toot-Trapper. But imagine if you were up in space in a space station or outside in a space suit. There are no windows. These earthly products were first developed for space travel, and are now incorporated in military uniforms for more deadly forms of gas. A napkin is also available for feminine needs.

I predict there will be many more products like these as people discover the potential and versatility of charcoal, not only to adsorb unwanted embarrassing personal odors, but for many other discomforts. If you are curious whether or not there is a product on the market for some special need, try doing a word search on the Internet. You may be surprised.

Chronic Intestinal Pseudo-Obstruction

Dr. Thrash shares two more heart-warming stories:

> "Several years ago, a middle-aged lady came to our conditioning center 'as a last resort'. She had had abdominal surgery at least twelve times in the previous five years, and was having to be admitted to the hospital every two or three weeks for gastric intubation and intravenous fluids. She was rapidly becoming debilitated. We put her on large amounts of activated charcoal by mouth, and placed fomentations to the abdomen several times a day. On the third day, she began to get abdominal distention and nausea. She was given a large dose of charcoal by mouth and a fomentation was applied, which was afterward replaced by a large charcoal poultice to the abdomen.
>
> "The following morning, the distention was gone, and we rejoiced that she was hungry. The treatments continued, and she was given a regular vegetarian diet. She never had another attack! She continues to use charcoal daily."

Pancreatitis

> "A twenty-six-year old woman came to the conditioning center with chronic relapsing pancreatitis, planning to stay for two or three months.

She had had several abdominal operations in the past three years, and had finally been diagnosed as having chronic pancreatitis. Because of the severe unremitting pain and debility, her physicians were considering total removal of her pancreas.

"Hoping to find some relief short of this drastic procedure, her husband had had his job transferred to our area so that she could be treated at our institute. They "knew" treatment would take several months. The patient was on potent medications for pain and nausea. Despite these, she was having continuous pain, nausea and vomiting. We asked if she were willing to stop her medications, which we felt were aggravating her condition. She reluctantly agreed.

"She was given fomentations to the abdomen, charcoal by mouth, and large abdominal poultices at night. Because of the severity of her symptoms, her lifestyle counselor stayed in her room continually for the first forty-eight hours. The following day, the nausea was better, and she could retain liquids. She was wheeled out into the sun for a few minutes several times a day. Day by day her pain declined, and her appetite improved. In two weeks she was taking long walks, and by three weeks was walking five to six miles a day. After four weeks, she was completely symptom-free, and they returned to their home, praising the Lord."[8]

Broken Bones

Joe is an older man in his late sixties. He stubbed his foot badly, resulting in some broken toes. They refused to heal. The doctor felt the only recourse was to amputate the foot if it did not soon make a change for the better. So he scheduled Joe for surgery in two weeks.

Joe was not prepared to lose his foot, but he was ready to try something, anything. Pastor John, an "evangelist" for charcoal whenever the opportunity arises, promptly introduced Joe to the wonders of carbon science, and gave him a quart jar of charcoal.

Pastor John's memories of charcoal go back to his boyhood. "My mother would nurse me and my siblings back to health with various home remedies. One of them she referred to as 'black magic' or 'black chocolate pudding.'"

Obviously the example stuck with Pastor John, and later when he and Sharon went to work in Ireland, the knowledge of charcoal's properties went with them. When they returned to America, he brought some homemade charcoal back with him. All these years later he still happened to have some left:

"That was twenty odd years ago, and that was the bulk of what I gave Joe. There was some twenty-year-old commercial grade charcoal mixed in too. I gave him instructions on how to put some water in a basin large enough for his foot, and stir in the charcoal powder, and just let his foot soak in that for an hour or more.

"Two weeks rolled by, and Joe arrived back at the doctor's office for an exam. They were incredulous when Joe handed them two fragments of bone that had erupted on their own from the damaged toes. He also showed off his foot that was well on its way to mending. 'Whatever you are doing keep it up,' was all the nurse could suggest. Joe is on his third quart now, but he is no longer worried about losing his foot.

"You can see that even though the charcoal was twenty years old, it obviously had not depreciated significantly in its adsorptive powers.

"I take it wherever I go. I pity those who still take it as a powder in water. It is chalky/gritty, unpleasant to swallow, and leaves my mouth, lips and teeth black. Not a good introduction. So I simply mix activated charcoal together with slippery elm powder four to one and leave it dry in a sealed jar. Then whenever I need it, it is instantly ready. I mix one tablespoon with enough water to make a thick jelly. This I can swallow easily, and it leaves no black moraine behind."

Thanks John. I'll put slippery elm powder on my list along with charcoal powder right now. CR

---

1 *Orthopaedic Medicine*, September 30, 1985

2 *Rx Charcoal*, p. 47, 48

3 Muck, O, *Behavior of Animal charcoal in Respect to Lesion Caused by Bacillus Pyocaneus*, Muenchener Medizinische Wochenschrift, No. 6, p. 297, February 8, 1910

4 Csillag, J, *Animal Charcoal in the Treatment of Eczema*, Dermatologische Wochenschrift, 95:1328-1329, 1932

5 Beckett, R, et al, Charcoal Cloth and Malodorous Wounds, *The Lancet*, p. 594, September 13, 1980

6 *Emergency Medicine*, September 30, 1985

7 Liebman, Bonnie, Who ya gonna call? GasBusters. *Nutrition Action Healthletter*, May, 2003

8 Thrash, *Rx: Charcoal*, pp. 75-77, 1988

Chapter 14 Super Natural

Chemistry is a fascinating science. It can be so exact. A plus B equals C. But when some drug is concocted to counter some known biological process in the body, besides producing the desired effect, scientists discover, more often than not, several very negative side effects. The more scientists study the biochemistry of man, what at first seems so straight forward, the more and more complex it is seen to be. What works in one instance may not work in the next. In addition to those anomalies, more and more diseases are becoming resistant to conventional drug remedies.

Considering the amount of junk food consumed by the typical American; considering the addictive beverages being drunk in place of water; considering how many pollute their lungs with smoke; considering the lack or poor quality of exercise the average person gets; considering the stressed pace most people maintain with too little quality rest; considering all these quite common practices that leave so many so very sick; and then considering the toxic nature of drugs, with their adverse side effects, that are ingested daily by millions trying to mask the symptoms of their self-induced diseases; one has to conclude most people are far more caring and gentle with the cold hard steel we call cars than with the human body. What is needed is something that works in harmony with the most amazing, technologically advanced, piece of engineering and design anywhere—our bodies. What is needed is something not so taxing upon the body's chemistry, something not so draining, not so depressing and debilitating on the body's own intricate and finely tuned healing mechanisms. What we should demand is a philosophy of healthcare that works to support the body's own restorative powers not the present system that so often promotes drugging and cutting and burning. It is time for a change. It is time that we abandon our dependence on the drugstore in favor of a more rational and intelligent practice of healing. What is needed is something that is tested, tried and time-proven to be harmless yet effective.

The different sections in this chapter focus on the positive results from using charcoal in various disease conditions. It will be shown to be effective when given internally, or externally as a poultice. Charcoal works just as consistently in the home as in the hospital or on a remote island.

Liver Cancer

Helen has seen her share of pain and death. From nursing in Viet Nam during the war to nursing her family and live-in patients here in America, she has seen the far reach of disease and suffering. She has seen what works in nursing people back to health, and what doesn't, and she understands our limitations. You accept the inevitable of some disease but you still work and pray to bring some measure of relief:

> "Two women who had cancer of the liver came to stay with us. The first woman was terminal. Her doctor had given up and offered her no hope. The liver was so distended that it interfered with her breathing as it pushed up against her lungs. You can imagine her anxiety. We radically changed her diet to a very simple cleansing program, and at the same time applied charcoal poultices over the liver for up to eight hours during the day and throughout the night. After just three days she was actually breathing easier and her liver was noticeably less distended. She was ecstatic. She sincerely felt that we had helped her, and she did survive a year beyond her doctor's dismal prognosis.
>
> "We put the second woman on a similar program, plus a full fever treatment to stimulate her immune system. Within a couple of weeks the swelling was much reduced, and brought marked relief to the woman."

Incredible results considering how simple the treatments were, but is it really possible that something on the outside can have so profound an effect on the inside?

As reported in *The Lancet*, research has shown that oxygen adheres to charcoal in hemoperfusion for temporary artificial liver support.[1] Researchers suggest this characteristic may be helpful in other treatments specifically intended to enhance oxygen supply to certain organs. Is it then just coincidence that other cases of severe liver failure show marked improvement when the blood is filtered through a charcoal bed? In fact patients with liver and kidney failure can sometimes be treated right in the home with large compresses placed over the back or stomach area. Taking charcoal by mouth will also help to prevent the build up of poisons that make the work of these organs more difficult.[2]

Liver Dialysis

Donald J. Hillebrand, MD, is associate professor of medicine, chief of hepatology, and medical director of liver transplantation at Loma Linda University Hospital. He reported, before the 53rd Annual Meeting of the American Association for the Study of Liver Diseases, that patients with episodic type C hepatic encephalopathy (EHE) may be able to benefit from dialysis using a charcoal-based liver dialysis unit (LDU). In a prospective study of eighteen patients with EHE, sixteen showed significant improvement in mental status within two days. In their study abstract, Dr. Hillebrand and colleagues write, "Charcoal-based hemodiabsorption utilizing LDU treatments are able to safely, rapidly and effectively resolve EHE failing to respond to twenty-four hours of appropriate medical management in patients with advanced cirrhosis."[3] This can represent a significant benefit to patients who otherwise can face a 10% to 30% risk of death. Other indications the Loma Linda team is considering for treatment with the LDU, include chronic liver failure, hepatorenal syndrome, liver failure after surgical interventions such as cholecystectomy, and liver transplant recipients who receive a marginal graft organ.

Coincidentally, no one knows why, charcoal taken by mouth has also been found to relieve the itching that is often associated with long-term dialysis.

Most likely you and I are not in a position to access this kind of technology directly. However, my experience twenty-six years ago with a man with liver cancer (Chapter 2) demonstrated conclusively to me, as have the experiences of others since then, that charcoal can work as effectively in the home as in the hospital, without the added stresses.

Unconscious

Several years ago, I conducted an evening course on home remedies for a community college. There was a keen interest in the demonstrations and anecdotal stories. But before the course ended one participant called to tell me that she could not make it to the rest of the classes. Her young son had been involved in a serious car accident, and was in a coma in the hospital in another city. The doctors had given the parents little hope of recovery. His liver and kidney functions were all critically high and none of their procedures were able to stabilize his condition. She asked if there was anything, anything at all I could suggest.

You can imagine the desperation this mother felt. Because the young man was unconscious, I suggested large charcoal poultices over the liver and over the kidneys. These were to be changed often. I said goodbye to the woman and told her we would lift them up in prayer. Under the circumstances I questioned what the doctors would allow.

I did not see or hear from the woman again for several months until we met at the supermarket. I was reluctant to ask about her son, but she confirmed that he had died. Then she added. "Remember what you suggested? Well, the doctors allowed us to apply the charcoal poultices, and within hours all his liver and kidney functions normalized. But the doctors felt that he had so much internal trauma that his body was not able to recover even with the improvement."

One is left with so many questions in experiences like these. But once more it was evident that charcoal was ready to do its part.

Liver Functions

Here are lab tests to demonstrate that what can happen when charcoal is taken internally may also be effected when applied externally.

In a letter to his doctor, Peter writes:

> "Following my liver biopsy test last January, you wrote me to advise that the biopsy was essentially normal, and that there was no further treatment needed at this time. While I appreciate such good news, I also took the advise of a close friend who suggested an old fashioned remedy that would expedite my liver returning to it's normal function and enzyme level. Since last February, I have been drinking a large glass of water every day that contains a large tablespoon of activated charcoal. Twice a week, for the first month or so, I also slept at night with a charcoal poultice taped over my liver. Last week I had my doctor in Mt. Vernon, draw some blood and run the liver tests again for me. The results were, I thought, truly amazing.
>
> - Alkaline Phosphatase—99 down from 218
> - GGTP—89 down from 302
> - SGPT—31 down from 229
>
> "During this same period all the symptoms of my hiatal hernia have disappeared, my cholesterol level has dropped from 293 to 270, and my triglycerides level from 299 to 260. I hope this update report is of interest and possibly assistance to you in treating similar patients."

One can only hope that Peter's doctor made good use of this clinical proof of charcoal's detoxifying and restorative powers. But whether he did or not, these benefits are ready to accrue to your health too just as soon as you put charcoal to work.

Jaundice In Infants

I mentioned early on about our experience with our firstborn. Because his liver was sluggish in filtering out the elevated bilirubin, Nathan was jaundiced when he was born. Out into the sun he went for a daily sunbath. The sunbaths, coupled together with charcoal slurry water given in a baby bottle, soon put Nathan back in the pink.

For infants, add one tablespoon of the charcoal powder into four ounces of water. This makes a good slurry that is able to pass through the nipple. Shake well before giving. Or, you can let the charcoal settle out, pour off the gray water and give that.

Doctor Agatha Thrash tells the following case of neonatal jaundice in a four-day old breast-fed baby:

> "The father took the baby to our laboratory to be tested for its total bilirubin levels. The levels continued to climb over the next twenty-four hours and a consulting physician agreed with our suspicion of an ABO blood incompatibility. When the bilirubin rose to 18 mg% the consultant prepared to give an exchange transfusion of blood.
>
> "The same hour the mother began administering as much charcoal as she could get the baby to accept. With the baby undressed in her lap, she sat in the sunlight giving over an hour of exposure to both front and back (babies can tolerate more sunlight before getting a sunburn than can adults).
>
> "At the next six-hour bilirubin check, the level was down to 16.5 mg%, and we knew we had avoided the hazardous exchange transfusion. Continuing with this treatment the bilirubin began to clear and was down to 4 mg% by the tenth day."[4]

In one astounding study the need for exchange transfusions in babies with erythroblastosis fetalis was cut by more than 90% with the use of charcoal.[5] Erythroblastosis fetalis is a severe anemia that develops in an unborn infant because the mother produces antibodies that attack the fetus' red blood cells. The antibodies are usually caused by Rh incompatibility between the mother's blood type and that of the fetus (that is, the mother and baby have different blood types).

These babies can be at extreme risk after birth and, depending on the severity, a blood transfusion may be performed. In one study done at Fort Benning, Georgia, activated charcoal, suspended in water, was given every two hours. The treatment was continued for 120 hours in normal newborns and 168 hours in premature infants, or until bilirubin levels fell. Charcoal should be begun at four hours of age to produce the maximum reduction in elevated bilirubin levels.[6]

Candidiasis

Richard Kaufman, PhD, is a bio-nutritional chemist. He reports, "Activated charcoal can be an effective adjunct to any regimen for the treatment of systemic Candida albicans infections." The toxins produced by Candida, absorbed by the blood and carried throughout the body, are effectively adsorbed by charcoal. Candida toxins cause allergic reactions and are responsible for the debilitating symptoms of candidiasis.

Charcoal curbs the growth of intestinal-based yeasts, and it counteracts the Herxheimer reaction. When, as a result of a successful treatment, there is a large die-off of yeast cells, there is a severe, short-term aggravation of Candida symptoms due to the increased amount of toxins produced. The Herxheimer reaction is often so unpleasant that patients abandon treatment before completion.

Taking activated charcoal is one method for alleviating the symptoms of yeast die-off so that patients can continue their treatment and not suffer. Dr. Kaufman suggests 20-35 grams of activated charcoal a day in divided dosages on an empty stomach until the problem is eliminated. The larger amount is taken for more severe situations. He recommends not taking charcoal within two hours of taking required medications.[7]

Post-Surgical Colic

After many experiences, Dr. Dana, NMD, has discovered that charcoal is an excellent preventive for what she terms "Killer Colic":

> "I often prescribe charcoal as a preventive for post-surgical colic. I recommend that those of my patients scheduled for intestinal or gynecological surgery take six capsules of activated charcoal at the second to last meal before stopping food intake in preparation for surgery. At midnight, they are to take more charcoal, and then nothing more at all until after their surgery. In my own case, after surgery, the doctors kept waiting for me to pass gas as a sign that my bowels were again functioning. They did not want to feed me or release me, but I assured them that I had no gas, and that my bowels were working just fine."

Hemorrhoids

Julie nurses at Wildwood Hospital:

> "A man arrived at the clinic with very painful and bleeding hemorrhoids. One was external and larger than half an inch and had a central ulcer. He had been suffering for a week. The doctor prescribed several things, including a Sitz bath, topical ice, and a mixture of various herbal teas applied externally as a douche. But, the man was convinced it was the charcoal paste that had brought him the most dramatic relief. The nurses mixed one teaspoon of activated charcoal powder with two tablespoons of olive oil to make a paste. This was applied to the rectal area overnight. Three days later the external hemorrhoid had significantly decreased in size, and there was no more pain. The bleeding had also all but stopped."

He returned home with instructions to continue the treatment. He was also instructed on how to improve his bowel health with regularity, increased water intake, exercise, and a diet higher in natural fiber.

After hearing this story, I decided to check out the Web. I was pleasantly surprised to find a relatively new product that apparently has been nationally advertised as a one-time application for relieving hemorrhoidal discomfort. The label reads: "Helps relieve these local symptoms associated with hemorrhoids: pain, itch, burning, soreness, discomfort."

The main active ingredient listed is activated charcoal (31.7%). It is described as a cream, but comes more like a paste and is used intrarectally, or can be applied externally. The website allowed people to write in and tell about their experiences. Of the thirteen who had responded, ten gave the product a five out of five score, and three gave it a one out of five score. The only objection those gave who did not like it, was that it was difficult to apply. Here are a couple of the positive remarks:

> "I agree with the FIVE star rating (and if there were six stars then I would give six). Not that I want to say anything negative about other brands, it's just that this one is in a league of its own. I have been using hemorrhoid products over and over again since I had my daughter and I have been waiting for a one-application treatment for so long. It's great to be able to think of something else than hemorrhoids for a while. The next time they come around, I have a date with [Product X]! Thanks!" —Kelly

> "I usually never take the time to write these things, but this is one time I just have to give you my opinion. I have never found a hemorrhoid product that comes CLOSE to [Product X]!! I had seen a TV commercial about it and so I thought I would try it, boy am I happy now! The itch and burning is gone, exactly like it says. This really is the ONLY one-application treatment. The next time I have hemorrhoids, I know where I will find my solution. Many thanks [Product X]." —Javier

But remember, as nurse Julie described, you can experiment making your own paste with charcoal powder and olive oil.

Prostatitis

The Doctors Thrash have found that charcoal has proven beneficial for inflammation of the prostate gland. They write in their book *More Natural Remedies*:

> "A hot [103°F] enema promotes healing. Use one cup of hot water and one tablespoon of powdered charcoal. Insert into the rectum with a bulb syringe or an enema setup. Allow to remain as long as possible, even overnight if it can be retained." [8]

Upset Stomach

Recently I talked with Yvonne. Some time ago she received a short pamphlet from me on charcoal, and decided to try it for her long history of upset stomach. She reported that she had gotten fast relief, and that she had had no more problems. She was preparing to go on a trip to Mexico, and was apprehensive about the reports of the Tourista Blues (diarrhea). I suggested that, in case of an emergency, she take some charcoal with her. Fortunately she only heard happy tunes, but she was thankful that she'd had a simple remedy she could trust.

Most cases of chest pain, called heartburn, are nothing more than acid indigestion that can be helped by taking charcoal. A slurry of charcoal and water, or a little olive oil and charcoal mixed together, is often all that is needed.

Pastor Emilio, now retired, was a popular public speaker. His speaking engagements involved traveling extensively, including overseas. He smiled as he told me:

> "I would catch every bug. If I noticed any symptoms of upset stomach or diarrhea, I immediately would take charcoal. Usually by the time my plane landed back in the U.S. my symptoms would be gone."

Ty is another world traveler. As a public speaker, director and field representative for a prominent publishing organization, he is often in the air on his way to some foreign country. With all his other important luggage to keep track of, Ty never forgets his emergency kit. "I take charcoal with me on all international trips, just in case. I visit in many different places, with different foods. If I detect the least upset stomach, I right away take some charcoal."

Sad to say, many continue to suffer needlessly. Little do they know, they have within easy access, a simple remedy they can safely carry wherever they go without the need of a prescription or the worry of Customs inspections.

In some individuals with poor bowel health, charcoal can be constipating. The solution is very simple. Increase the intake of water, cut the dosage of charcoal in half, or mix a half-teaspoon of charcoal with a half-teaspoon of Psyllium seed husks or flax seed meal together with eight ounces of water.

Bowel Disease

Marilyn always wore a pleasant smile and no one would have guessed that she had Ileitis. This chronic bowel condition is akin to Crohn's disease or severe ulcerative colitis. While the disease mechanisms differ, the suffering is common to all. Like other chronic bowel irritations, diet and stress are major triggers of diarrhea. But whatever the trigger, charcoal is able to alleviate the symptoms and the distress. Marilyn reports she has been able to control her Ileitis primarily by being careful with what she eats. But if it begins to give her problems, then, at the first sign of discomfort, she takes some charcoal capsules. Invariably they bring her prompt relief.

Here are further recommendations from the Doctors Thrash from their book *More Natural Remedies*:

> "Charcoal may be helpful in the control of diarrhea associated with Crohn's disease, colitis or irritable bowel syndrome, etc. Four to six tablets may be taken two to three times a day between meals. If the charcoal seems to irritate the colon, one to three tablespoons of powdered charcoal may be stirred into a glass of water, the charcoal allowed to settle out, and the clear water drunk".[9]

As well, bowel inflammation may be treated with a charcoal compress made with strong hops tea instead of water. Apply the compress at bedtime, and leave it on all night. Drinking charcoal slurry water, three or four glasses a day, is often very helpful. [10] This mixture may be put in a quart jar and sealed and used later. It may also be given in an infant's baby bottle in cases of colic.

Some physicians have used charcoal to calm spastic colons.[11] Even in the late 1800s physicians noticed that bleeding within the intestinal tract could often be checked with charcoal.

Diarrhea

Diarrhea, which accompanies many different diseases, may also be triggered by food poisoning, bacteria, nervousness and other factors. Whatever the trigger, symptoms of diarrhea may often be remedied with charcoal. One doctor suggests charcoal as a corollary treatment for the diarrhea that comes with HIV/AIDS.

John was desperately ill. Beth wakened to find him on the toilet. She had no idea how long he had been there. The floor was wet from his dripping sweat. He couldn't stand, and he felt like another bout of diarrhea every time he tried. John was running a high fever and was exhausted.

It was two in the morning, and Beth didn't know what to do. They had used charcoal so many times, but this seemed far more serious. It was all Beth could think of, so she had him take ten tablets as she handed him one at a time. Then, unsure what to do next, Beth prayed, "Lord I have done what I could. Now you need to bless this simple treatment." Within ten minutes, John was able to go to bed and rest. By morning the fever had subsided, and he improved steadily.

When John and Beth moved to India, they were, like most, plagued with dysentery. Typically, people were given a broad-based antibiotic containing Flagil. But because their daughter was allergic to the antibiotic, Beth only used charcoal for her.

Being surrounded by poor sanitary conditions, they were constantly fighting the diarrhea. In the first three years they finished off five thousand tablets. They also used the charcoal ground up as a poultice to put on boils and the many bites that easily became infected.

While it is not always possible to identify what the underlying disease may be, because charcoal is harmless, you may safely give it for any number of bowel problems. As mentioned above, those with especially sensitive bowels, as in chronic ulcerative colitis, can simply stir a tablespoon of charcoal into a glass of water, let the sediment settle out, and drink the gray water. Do this regularly with your daily intake of water.

Dr. Weis confirms, "Some of my patients with ulcerative colitis have also found that taking activated charcoal along with remedies such as aloe vera juice, acidophilus and psyllium helps keep their symptoms under control."[12]

When it comes to trying some new remedy, we all feel better if someone who is well qualified will give it their thumbs up. Mervyn G. Hardinge, MD, Dr. PH, PhD, the founding dean of the School of Public Health at Loma Linda University, has broad-based experience in the fields of pharmacology, nutrition, and health. Holding degrees from Harvard and Stanford, he has authored more than fifty scientific papers in peer-reviewed journals as well as several books, including the three-volume *Family Medical Guide*. We would expect he is eminently capable of speaking on the merits of charcoal. In his most recent book, *Drugs, Herbs, and Natural Remedies*, he places charcoal under the heading of "harmless". Of its more common uses, he lists it for relief from gas, as a laxative, for inflammation of the bowels, for colic, diarrhea, ulcers and pain.[13]

When I called and talked to him on the phone he assured me that he has always been a strong supporter of simple remedies. When I asked if he had any memorable charcoal stories he replied:

> "Yes. An old but very vivid one comes to mind. While attending a camp meeting, our fifteen month-old son contracted a good case of diarrhea. He was, nevertheless, still very active, running around. But as he ran the contents of his diaper dribbled out behind him making no little mess for us to clean up. By the time we were ready to leave, his bottom was so raw, he would scream when his mother tried to change his diapers. Driving through Oklahoma I stopped at several pharmacies along the way to see if they had any charcoal, but none of them carried it in their store.
>
> "Finally I stopped in a small village and was able to find some charcoal tablets. We ground them up, but then wondered how we were going to get a fifteen month-old baby to take it. We decided on peanut butter. We mixed it in with the charcoal, and within a short time he was no longer bothered

when it came time to change his diapers. I have always encouraged the use of natural remedies and we have used charcoal through the years for an assortment of infections. For instance, for a cut on the hand we would make a simple poultice and wrap the hand completely."

When I related this story as it is written, I was asked, "Did they give the charcoal-peanut butter combination to their son to eat or did they spread it over his bottom?" Let me test the reader, which would you have done?

If knowledge is a qualification for endorsing some natural remedy then we can surely trust Dr. Hardinge's recommendation of charcoal. When it is backed up with years of experimentation, we too can take comfort in knowing we are not being rash or presumptuous. Charcoal is more than natural, it is super natural.

Cholera

Diarrhea. Most folks look at this affliction as a disease, but initially, it is simply the body's way of trying to throw off some toxin or pathogen before they can produce even more serious complications. Now diarrhea is no little problem. If it does its job and the bowels return to normal function, then we can be thankful. But too often the mechanisms that initiate diarrhea do not shut off, and the problem continues to the point of severe dehydration and loss of essential electrolytes. In severe cases, death can quickly follow.

Diarrhea, along with vomiting, are common to many diseases and conditions, but they are classic symptoms of cholera. Cholera has not been common in North America for the last hundred years, but it is once more a concern. Because diarrhea and vomiting are also common to the various forms of food poisoning, it may be helpful to use cholera as a model for treating these various conditions.

In the last decade cholera reached epidemic proportions in South America. The cholera epidemic in Africa has lasted more than twenty years. A few cases occur yearly in North America among people who travel to South America or eat contaminated food brought back by travelers. While cholera is rare in industrialized nations, the disease is still common in other parts of the world, including the Indian subcontinent and sub-Saharan Africa.

Although cholera can be life threatening, it is easily prevented and treated. Because of advanced water and sanitation systems, cholera is not a major threat in North America. However, everyone, especially travelers, should be aware of how the disease is transmitted and what can be done to prevent it.

Cholera is a diarrheal illness caused by infection of the intestine with the bacterium *Vibrio cholerae*. The infection is often mild or without symptoms, but it can sometimes be severe. Approximately one in twenty of those infected has the severe form of the disease, which is characterized by profuse watery diarrhea, vomiting, and leg cramps. In these persons rapid loss of body fluids leads to dehydration and shock. Without treatment death can occur within hours.

A person may get cholera by drinking water or eating food that has been contaminated with the cholera bacterium. In an epidemic, the source of the contamination is usually the feces of an infected person. The disease can spread rapidly in areas with inadequate treatment of sewage and drinking water.

The World Health Organization (WHO) claims waterborne gastrointestinal infections cause eighty percent of all disease worldwide, and kill more than 50,000 people every day, over 5,000 from diarrhea and cholera. As a traveler, if you believe you are the exception, you are not. Up to fifty percent of all holidaymakers who travel abroad get diarrhea.

While working in Nepal, we were all made aware that as the rainy season approached, the number of cases of diphtheria, typhoid and cholera would climb.

Early in the book, I mentioned a case of cholera on a Pacific island that was successfully treated with charcoal. A year before, shortly after I left Nepal, an acquaintance died of the disease. There were differences and similarities in the two cases. The thing in common was the poor sanitation. The difference was the form of treatment.

In both countries, a lack of proper bathroom facilities and the resulting habit of defecating out of doors, without burying the stool, is the perfect formula for the rapid spread of disease. When the rainy season arrives, the run off from the land carries the different microbes into the streams and water supplies. From there, the waters quickly infect dozens or thousands of people.

This may sound incredible to many Americans, accustomed as we are to improved methods of hygiene and sanitation, but this is the all too common scenario in most developing countries. Even so, America is not exempt from the potential of epidemics. News reports regularly tell of the contamination of public water supplies, with hundreds of people being infected with some pathogen, and sometimes deaths follow. Prevention, namely proper attention to sanitation, is the first line of defense.

The cholera bacterium may also live in brackish rivers and coastal waters. Several people in the United States have contracted cholera after eating raw or undercooked shellfish from the Gulf of Mexico. While cholera was prevalent in North America in the 1800s, it has virtually been eliminated through the use of modern sewage and water treatment systems. However, as a result of improved transportation, more and more people from America and other western countries are traveling to parts of Latin America, Africa, or Asia, where epidemic cholera is occurring. Common sense recommendations for those traveling abroad are posted by the various Centers for Disease Control. A simple rule of thumb is "Boil it, cook it, peel it, or forget it."

The dehydration that results from the diarrhea that comes with cholera can be simply and successfully treated with an oral rehydration solution that replaces the loss of fluid and salts. This prepackaged mixture of sugar and salts is used throughout the world to treat severe dehydration. It is mixed with water and drunk in large amounts. Severe cases also require intravenous fluid replacement. With prompt rehydration, fewer than 1% of cholera patients die.

However, it is those 1% we do well not to ignore. The government health worker I talked to on a remote Pacific island realized that her standard treatment for cholera was not working. It was not until she turned to charcoal to control the diarrhea, so the rehydration solution could do its work, that her patient recovered. In that severe case of diarrhea from cholera, charcoal was the lifesaver.

How much charcoal should be given? You cannot give too much, but two heaping teaspoons of pulverized charcoal four times daily should be adequate. If the patient is unable to drink adequate fluids to prevent the severe dehydration that characterizes cholera, supplement the charcoal with one to two and a half liters of normal saline intravenously.[14]

It is also worth remembering that as a preventive, charcoal effectively adsorbs the toxins that are released on infection from diphtheria, tetanus, and dysentery bacteria. In one study a 10% solution of diphtheria toxin was exposed to charcoal for one hour. The clear liquid was then injected under the skin of a guinea pig. There were no adverse effects at all. But the control animal died within forty-eight hours when give only 1% as much toxin.

Similar results were obtained with tetanus and dysentery toxins. Researchers have also demonstrated unmistakable evidence that activated charcoal prevents the expected toxic effects of endotoxins normally released by *Vibrio cholerae* and *Escherica coli*.[15] Once again, science has verified in the lab what the eclectic health worker has already discovered in the field. Namely, that charcoal, in conjunction with established nursing practices, should be, in certain cases, a preferred remedy in emergency or primary care.

We have read how charcoal worked to save the life of an infected guinea pig in a research lab. We have seen how quickly and effectively charcoal can work in the much more difficult later stages of cholera, even when far from a clinic. We have listened to stories of severe dysentery and food poisoning responding quickly to charcoal when taken orally and when applied as a poultice. We now have sufficient information to expect charcoal to work just as well for us and for our children in similar emergencies.

Typhoid Fever

Sandra travels back and forth between India and Nepal giving public health lectures, and operates an adult training school. Anyone who has been in that part of the world understands that serious illness is never very far away. And they know how easily the teacher can become the patient. I e-mailed Sandra and asked if, in addition to her experiences with scorpions, she had ever succumbed to cholera.

She wrote back:

> "I have not had cholera but I have had typhoid fever and bloody bacterial dysentery. For the dysentery I took a spoon of activated charcoal powder in a glass of water after each loose stool. I also ate five cloves of raw garlic with food three times a day. And I used a quality nutritional supplement. With this treatment I was well within a couple of days with no reoccurrence.
>
> "I also had typhoid fever a few years back. I used charcoal for the diarrhea part of it, and found steam baths the most helpful in combating the fever. Every evening as the fever would begin to rise, I would hit it with a steam bath. As long as I continued this, I managed to keep it under control.
>
> "Because of having to work full time, I was not able to get the needed rest and time for daily treatments, so it took me over three weeks to fully recover. With the steam treatments and charcoal, I also used garlic and a nutritional supplement. And of course, much prayer, and rest whenever I could. I was able to fully recover without the use of any antibiotics."

Prevention

Jon writes:

> "For several years I worked in a hospital emergency room, and one curious thing I observed, was something that few people are aware of. Whenever an overdose patient was brought into the emergency room, the very first thing a doctor would do was make the overdose patient drink activated charcoal mixed with water. During those years of seeing many overdose patients come into the ER, I never saw an overdose patient die who was conscious and who was able to drink the charcoal slurry. Because of my experience in the ER, I have, over the years, done some research on activated charcoal and discovered many beneficial properties. Some of these seem to border on the miraculous, and that makes me wonder why the benefits of it are not more well known."

As a result of his studies Jon began to experiment on himself. Today he is confident that charcoal can be just as effective as a preventive as it is as a remedy.

Jon posted his personal observation on his web page:

> "Influenza, 'the flu', is a highly contagious viral infection of the upper respiratory tract, and what is commonly called 'stomach flu' is technically

not really the flu, but rather, in most cases, gastroenteritis, which is an acute inflammation of the lining of the stomach. I thought I would share with you something that has helped me greatly over the years whenever I have experienced flu-like symptoms, such as when my stomach starts that unusual gurgling sound that precedes nausea, vomiting, diarrhea, chills, and fever. It is especially helpful if you work with a lot of people, up close and personal as I do.

"I have had the beginning of several of these symptoms over the past few years and the very first thing I do is get out my jar of activated charcoal powder. You may laugh at this or find it repugnant because it sounds too simple, or because the powdery, gritty, black particles make you think it is intolerable. Or maybe you find it can be a little too messy. But believe me, activated charcoal works wonders for many conditions. It's worked for me many times, and I've even been an eyewitness to it saving the lives of many people."

Hopefully, you never find yourself faced with any of the above emergencies, but if you do, you should be confident by now that charcoal can work just as super naturally for you as it has for others. CR

---

[1] Chamuleau, R.A.F.M., and Dupont, A, et al, Activated Charcoal and Ammonium Production, *The Lancet*, pp. 633-634, September 19, 1981

[2] *Ibid.*, 1:1301, 1974

[3] Hillebrand, Donald J, et al, AASLD 53rd Annual Meeting: Abstract 100249, reported in *Medscape Medical News*, Nov. 2, 2002

[4] Thrash, *Home Remedies*, New Lifestyle Books, p. 150-151, 1981

[5] *Medical World News*, February 17, 1967

[6] *Clinical Pediatrics* 4 (3) 178-180, March 1965

[7] Kaufman, Richard C. PhD, The Universal Antidote and Detoxifier That Extends Life: Activated Charcoal, *Journal of the MegaHealth Society*, July 1989

[8] Austin, Phylis, Thrash, Agatha, MD, & Calvin, MD, *More Natural Remedies*, Thrash Publications, p. 91

[9] *Ibid.*, p. 43

[10] *Ibid.*, p. 122

[11] *Ibid.*

[12] Drweil.com

[13] Hardinge, Mervyn G, MD, *A Physician Explains Ellen White's Counsel on Drugs, Herbs, & Natural Remedies,* Review & Herald Pub. Assoc., pp. 161, 164, 2001

[14] Thrash, *Home Remedies*, New Lifestyle Books, p. 145, 1981

[15] Cooney, David O, *Activated Charcoal*, TEACH Services Inc., New York, p. 53, 1999

Chapter 15 Camels, Puppies, Chickens, and Other Friends

What about our four and two-footed friends? Could their biochemistry and physiology also benefit from the wonders of charcoal? Just because you haven't seen it on the shelf at your local animal clinic or country co-op, doesn't mean it won't work for animals or that many informed vets aren't already using it. It probably means you just haven't thought about asking them to stock it.

One Sick Camel

Alexandra and Lew made a seven-year, 17,000-kilometer expedition around Australia with a camel train. During their trip, they were always concerned for their camels, what with so many poisonous reptiles and plants, and poisoned bait set out for dingos and foxes. At one stop, they released their camels within a local stockman's backyard corral—the last place one might expect to find a poisonous plant. Alexandra writes:

> "It was an innocuous solitary shrubby thing that was growing slap bang in the middle of the enclosure. The yard owner had said it was called a 'Bullock bush' and that animals used it for shade and he had left it at that.
>
> "It was only four or five hours after being released in the yard, when I saw Burke, the Nomads camel, who is a fairly bombastic individual, couched by the gate. All the others were milling around elsewhere. I sensed something might be amiss. Half an hour later he had moved position but was still couched.
>
> "A quick scout of the area revealed that one of the camels had obviously decided to try out, albeit just a nibble, the 'Bullock bush.' We surmised it must have been Burke, who is a notorious greedy-guts. There were no vets for hundreds of kilometers, and as it was late in the evening we decided to be on the safe side and started treating Burke as if it were a case of poisoning.
>
> "Drenching a camel that wants nothing more than to be left alone to wallow in his tummy ache is not an easy task. For a drench, we came up with one of the big drink bottles from the bum bags. We managed to pull out the little plug and clamped a meter length of hose onto the remaining nozzle.

"Presto! A drench bottle, which proved extremely effective despite Burke's protestations. In this way we were able to administer 250g of activated charcoal in powdered form. We also injected the camel with Vitamin C, as it is widely believed to help combat the effects of toxicity.

"That night I slept fitfully and checked on Burke four or five times. But, by morning it was clear he was not improving. It was now imperative that we obtain a proper plant identification. An old boy in the service station bar took the plant cutting from me, looked at it and mumbled, "any stock that eats it, dies". Now the cold realization hit, if my camel was to have any chance at all, time was of the utmost essence

"We discovered the name of the plant was Boubillia. A phone call to the Port Augusta vet confirmed the old man's edict—the plant was highly toxic and there was no known antidote. Burke's prognosis was grim, and the vet suggested I prepare myself for the worst. The only other suggestion he had was to try and flush the poison out of the camel's system with liquid paraffin. We were looking at a 400km round trip to Port Augusta if we had any chance of saving the red camel with no hump.

"Before leaving for Port Augusta, we drenched Burke again with a further 250g of activated charcoal suspended in about three liters of water, and gave him another shot of Vitamin C. On our return seven hours later, we drenched him with the liquid paraffin mixed with more activated charcoal, and also gave him an injection of Vitamin B complex.

"Again, that night I checked him several times. The following morning, Tuesday, we repeated the drench this time suspending the activated charcoal in the liquid paraffin, and once more hit him with the Vitamin C and B complex. Our spirits lifted when he passed some very black manure and began to chew his cud occasionally.

"On Thursday, much to our relief, he stood up, wandered around a bit, and showed no signs of sitting down again. We felt quietly confident that Burke was over the worst. What had been intended as a one-night camp in Pimba had ended up as a five-day delay while we treated Burke. The owner of the yard was amazed that Burke had even tried to eat the bush, whereupon he explained to us that the reason the stock in the area used the bush for shade was because they knew instinctively that it was poisonous. Now he tells us!

"As to whether Burke would have recovered without our intervention? For what it's worth, my own feeling is that our actions probably did contribute to Burke's recovery, particularly as it was caught so quickly."[1]

Alex e-mailed me at the end of their seven-year journey with this epilogue:

> "Activated charcoal ended up savin' the day on another occasion when we were taking the camels to Cape York. One camel ate a mouthful (that's all it takes) of Cooktown Iron Wood. Jaws took four days to recover, but recover he did."

With all poisonings, time is of the essence. But animal studies have shown that when it comes to charcoal as an antidote, quantity definitely is a factor. A mature camel can weigh up to 700 Kg (1542 lbs). Minimum suggested quantities of charcoal for livestock poisoning begin at 1 to 1.5 grams of activated charcoal per 1 kilogram of body weight. If Burke weighed in at 600 Kg then you can see he may have responded much quicker if Alex had used 600—900 grams of charcoal instead of just 250 grams.

Bitterweed

In three separate trials, researchers looked at the potential of activated charcoal to relieve the effects of bitterweed poisoning. The first group of lambs was offered very small amounts of bitterweed and varying amounts of charcoal according to their body weight. After ten days the lambs simply refused the bitterweed. In the second group, lambs were dosed with bitterweed and then fed grain with varying levels of activated charcoal. A decline in their feed intake indicated they were experiencing the poisonous affect of the bitterweed. Those lambs that received charcoal ate more feed than those that did not receive charcoal. The third group of lambs was fed a food supplement with or without activated charcoal and then exposed to bitterweed. As we would expect by now, those lambs supplemented with activated charcoal took more bites of bitterweed than lambs not receiving charcoal. The lambs showed no aversion to eating charcoal mixed with their food.

Together, these results suggest that charcoal somehow diffused the poisonous effects of bitterweed even when the lambs continued eating it, and charcoal should be considered as a supplement to stock feed.[2] What with poisonous plants, mad cow disease and other toxic organisms, charcoal should look better and better to the livestock industry every day.

Monkeys

Animal husbandry is becoming more and more active in exploring and marketing charcoal products for a range of different ailments, but some animals just can't wait for the wheels of progress to catch up with them.

Furry jungle inhabitants have more to contend with than just parasites and microbes. Some of the most nutritious plants that they eat also contain more or less toxic substances called secondary compounds. These compounds act as a defense mechanism against hungry herbivores. Red colobus monkeys on Zanzibar Island, Tanzania, prefer leaves of the exotic Indian almond and mango trees. These trees yield leaves high in protein as well as secondary compounds called phenols, which interfere with the monkeys' digestion.

What could these animals eat to counteract the poisonous nature of the leaves, while retaining their nutritional benefits? For six years, anthropologist Thomas Struhsaker, of Duke University, studied the fascinating feeding behavior of the Tanzanian red colobus.[3] Besides having a preference for almond and mango leaves they also eat charcoal from charred stumps, logs, and branches, as well as from around man-made kilns.

"They really go after the charcoal. Bigger monkeys try to take charcoal away from smaller ones. And they come down from the trees to grab pieces much bigger than they can possibly eat, carrying it off with two hands." A colleague, none other than University of Wyoming chemist, David Cooney, showed that the charcoal had a high adsorptive capacity for phenols. But while the toxic phenols adhered to the charcoal, the proteins did not. Interestingly, birth rates and population densities of the red colobus are significantly higher where charcoal is found in conjunction with almond and mango trees, than where there is no charcoal.

The Right Antidote

Beth's girlfriend had a dog that ate rat poison. What could she do? She gave it charcoal and, not surprisingly, it quickly recovered.

Now, animals can be a bit like people when it comes to charcoal. At first they have no inclination to try it. So, just mix the powder right into their food. You can also quickly make up some oatmeal and add charcoal powder along with chili powder, garlic, salt or any other spice to make it smell a little more like meat.

Eating Easter lilies and day lilies can leave your cats very sick. Again, early treatment is the answer. Give charcoal at a ratio of 1-3 grams per 1Kg of body weight and lots and lots of water for forty-eight hours.[4] On their website the ASPCA includes it as an antidote for rat poison, especially for bromethalin.[5] Pets commonly ingest aspirin, and charcoal works every bit as well with your favorite Buddy as it does for Johnny.

In fact, one pet site on the Internet also included charcoal for treating Tylenol, antifreeze, pesticides (pyrethrins and pyrethyoids), and insecticides. If you detect the poisoning within four hours, you will have a quicker response to charcoal. However, if rushing to a vet is not practical, and if the pet is still conscious, immediately start pouring a heavy mixture of charcoal slurry water down their throat.

Animal Poison Control Center—USA 1-888-426-4435

Here are the recommendations on how to manage a poison emergency posted by the Animal Poison Control Center:

- The recommended dose of activated charcoal for all species of animals is 1-3 gm of charcoal per 1 kg body weight.
- Repeated doses of activated charcoal every four to eight hours at half the original dose may be indicated when there is a possibility of reabsorption of poisons filtered out by the liver.
- Activated charcoal can be given orally with a large syringe or with a stomach tube.
- Activated charcoal should not be given to animals that have ingested caustic materials.

It should be noted, that those substances that are only slightly adsorbed by charcoal in humans, are likewise poorly adsorbed in your pets. These substances include ethanol, methanol, fertilizer, fluoride, petroleum distillates, most heavy metals, iodides, nitrate, nitrites, sodium chloride, and chlorate. But as with Burke the camel, if you have nothing else charcoal will not hurt. Again, as you would with any liquid, avoid having the animal breath in any of the mixture. This is especially true when the animal is unconscious or uncooperative.[6]

Worm Medicine

Dr. Means is a veterinary toxicologist at the ASPCA Animal Poison Control Center (APCC) in Urbana, IL. She tells the story of a pet owner and her eight-year-old Lab retriever who loved to 'help' his owner feed the hoses. On one occasion his owner inadvertently dropped a slice of apple spread with ivermectin, a paste wormer for hoses, and Martin grabbed and ate it. The owner knew that others used the same drug to treat their dogs for heartworm so she wasn't concerned. But, when she later discovered Martin staggering like he was drunk with no interest in his surrounding, she quickly loaded him up and headed for the vet's office.

Martin was given activated charcoal. And, he continued to receive multiple doses of activated charcoal, because ivermectin persists in the body for several days and keeps returning to the gut before being excreted in the feces. Martin also received intravenous fluids. Martin was discharged four days later, but some dogs require several weeks to recover.[7]

Incidentally, Martin's experience with ivermectin poisoning is not unusual. It has been noted most often in dogs following overzealous treatment with an ivermectin-containing product formulated for horses or cattle. The breeds of dogs most commonly affected are collies and collie-crosses.

Grapes and Raisins

Dr. Means tells about another big, playful Labrador retriever who often got himself into some sticky situations. One day Magoo snagged a box of raisins from the pantry, and ended up eating an entire pound of the sweet treats. Other than losing their treats, Magoo's owners didn't think much about it. Lots of people share grapes with their dogs, and often use raisins as training rewards. So, it hardly seemed the kind of emergency that required a call to the veterinarian.

Nevertheless they did call and found out that around 1999, the APCC began noticing a trend in dogs that had eaten grapes or raisins. Nearly all developed acute kidney failure. In all of the cases, the ingredients for potential acute renal failure were the same. It didn't seem to matter whether the grapes were purchased fresh from grocery stores, grown in private yards, or which brand was eaten. The amounts ranged from a single serving to over a pound, and the cases were evenly spread geographically. Dogs who ate the grapes or raisins typically vomited within a few hours of ingestion. At this point, some dogs would stop eating, and develop diarrhea. The dogs often became quiet and lethargic, and showed signs of abdominal pain. These clinical signs lasted for several days, sometimes even weeks.

As kidney damage developed, the dogs produced less and less urine. Death occurred when they could no longer produce urine. In some cases, dogs that received timely veterinary care still had to be euthanized. To date, after extensive studies, there is still no known reason why the animals reacted as they did. But, even though the exact cause of the kidney failure is unknown, dogs can be treated successfully by inducing vomiting soon after the ingestion of grapes, and giving activated charcoal to help prevent absorption of potential toxins. It is also recommended that dogs should be hospitalized and placed on intravenous fluids for a minimum of forty-eight hours.

Slug Bait

Another unsuspecting but common cause of poisoning in pets is snail and slug bait. Metaldehyde is a common ingredient in commercial snail and slug baits in the United States. These products are often used around gardens in the southern United States, Pacific coast, and Hawaiian Islands where snails and slugs are more prevalent. It comes in commercial liquid, powder, and pellet products. Various trade names include Snarol, Buggetta, Deadline, Slug Death, Slugit Pellets, Slug Pellets, Mini Slug Pellets, Namekil, and Optimol. Although less common, metaldehyde baits are also formulated with other chemicals such as carbaryl or arsenic. The bran or molasses sometimes added to the bait makes it more attractive to snails and slugs, but it also attracts domestic animals. Animals that ingest metaldehyde, which is a neurotoxicant, may experience vomiting, rapid heartbeat, tachypnea, staggering, tremors, and seizures. Death can also occur. It is less likely in cats but any dose in dogs warrants treatment.

In dogs and cats, activated charcoal administration at 1 to 4g per 1kg body weight is recommended. Repeated doses given at half the original dose every six to eight hours may be of benefit as well. Signs of poisoning in dogs may begin after a few minutes or up to three hours after ingestion. Other signs may include anxiety, panting, salivation, vomiting, diarrhea, tremors, continuous seizures, metabolic acidosis, rigidity, and high temperatures. Delayed signs that may develop are depression and coma. Death from respiratory failure can occur within a few hours of exposure. Liver failure may develop two or three days after exposure.[8]

Black Walnut

Not only do we live in a world of man made poisons, but there are the natural kinds too. Take for instance black walnut. Horses that are exposed to the shavings or sawdust from walnut trees are most at risk because it is occasionally used as animal bedding. A compound known as jugalone has been suspected to be the toxin, but efforts to document this have been inconclusive. Signs in horses occur within twenty-four hours of exposure to walnut shavings, and include rapid onset of laminitis, an increase in pulse, swelling of the limbs, rapid breathing, and elevated temperature. Necrosis of the dorsal laminae may occur and complicate recovery.

Horse and Farm magazine recommends the source of the walnut should be removed, and gastrointestinal detoxification carried out using mineral oil or activated charcoal and a mild cathartic. The legs and feet should be washed.[9]

Laminitis Antidote

Laminitis is an inflammation of the sensitive plates of tissue in a hoof, especially a horse's hoof, usually causing lameness. Since so many triggers of laminitis end up creating toxins in the gut, something that adsorbs and neutralizes toxins in the gut makes perfect sense. Did your horse eat something poisonous or did your horse founder? The product Universal Animal Antidote Gel (UAA Gel) is made of activated charcoal and clay. It has been used with good effect on acute laminitis. Sometimes one dose is all that is needed. The only caution is, avoid over-doing, as it can have a constipating effect. Because it has little taste, there is less resistance to giving it. Another product containing activated charcoal claims, not surprisingly, to remove objectionable odors from foul or infected wounds, to be good for slow-healing sores and infected lesions, to aide in the prevention of proud flesh, and can be used on wounds, cuts, abrasions and capillary bleeding, with or without a bandage. There are now several products on the market designed specifically for livestock and pets.

Animal Feed

As mentioned earlier, Aflatoxin B_1 and T-2 toxin are both fungal toxins that kill both humans and animals. The fungi associated with these toxins often bloom in grain elevators and often end up in cattle or chicken feed. Charcoal is a powerful antidote to these and other fungal toxins. Algae also produce their share of toxins. It has been observed that animals that drink stagnant water containing a certain class of blue-green algae quite commonly "die right on the banks of the pond". The toxins, like cyanide, act very quickly. They are so deadly, the only way charcoal can be effective is if it is given **before** swallowing these poisons or **immediately** after.[10]

Other Plants

Charcoal has also been recommended for poisoning from milkweed and Japanese Yew. And it is reported that charcoal has been successfully administered to sheep and cattle for lantana poisoning.[11] It would seem that animal husbandry is far ahead in using charcoal as a medicinal treatment.

If an animal is suspected of having had recent exposure to a toxic plant or chemical, treatment with activated charcoal is indicated. Activated charcoal is used to adsorb or bind up the toxic compounds before they are absorbed by the intestine. To minimize absorption by the intestine, some veterinarians recommend administering mineral oil to speed intestinal transit time. Activated charcoal and mineral oil are generally effective in all exposures to toxic plants or compounds but must be given as quickly as possible after ingestion.[12]

Home Nursing

After nursing all day at the hospital, Julie returns to her home to practice her veterinarian skills. After recommending charcoal for patients she returns home to use charcoal for her dogs and sheep. When her dog developed an eye infection, she applied a charcoal poultice over the eye. She changed the poultice three times a day for three days, and by the third day the infection was gone.

For her kitten's diarrhea, she gave a syringe full of charcoal water down her throat. This worked immediately. Then her sheep developed scours, and she gave it water to drink with charcoal mixed in. It too was soon well again.

Hoof and Mouth

In his smaller book, *Activated Charcoal,* David Cooney uses the last chapter to quickly review the more unusual medical uses of charcoal. It is very enlightening reading. Hoof and mouth virus, together with some of its cousins, have recently decimated herds of animals worldwide. These viruses can be entirely adsorbed on charcoal. Although charcoal does not destroy the virus, it does greatly reduce its activity. What is more, injections of the virus-charcoal complex were found to create immunity to the disease. It was also demonstrated that certain bacteria, while not destroyed, were effectively bound by charcoal, and their physiological functions were sharply reduced.[13]

Parvovirus

Joyce's neighbors had several dogs. One puppy was a favorite of the father's, but they were struggling financially, and had not had the funds to get the puppy its shots. The pup had contracted parvo (parvovirus). Parvo typically produces a noticeable smell, but the poor pup had a much more putrefying odor. The prognosis is not good. Most often at this stage the animal dies.

When asked by her neighbor what she thought they could do for the pup, Joyce confessed that she hadn't the faintest idea how to treat parvo. However, she did know that charcoal helps with bad odors from wounds, so she suggested they try some charcoal. She told the lady that she had no idea at all if it would work, but it certainly would do no harm. At least it would help eliminate the bad odor.

Joyce gave the lady some charcoal but when she mixed it with water the dog could not be induced to take it. So, she then mixed it with some prepared food and the puppy accepted that. The woman gave the pup another spoonful during the night, and the next morning the puppy was completely recovered. They had no further problems.

Wounds

Here we have an assortment of different animals and their variety of wounds. From bites to cuts, and from boils to injuries, they all became infected. Even after vets had given up, charcoal worked tirelessly on, and brought full recovery.

While I was in Nepal, Steve approached me and asked if I would go with him to visit a nearby farmer. When we arrived, we found that some dogs had attacked one of his prize goats. Some of the puncture wounds on the rump had become infected. There was a terrible smell of rotting flesh!

I had brought the scissors I used to trim my hair, and reluctantly used them to clean the hair off all around the wounds. We cleaned the area, and then applied a charcoal poultice. We didn't have duct tape and I can't remember how we were able to secure it, but it seemed to hold.

I didn't go back to see the goat, but Steve did go a couple times to clean and redress the wound. Within days it had healed over. The farmer was most grateful. Oh yes, if you are wondering, I did sterilize the scissors.

Not long after Jon and Bridgett were married, they took in a stray cat with a nasty open sore on its neck. They began treating it with charcoal poultices, and it soon healed. One pet site on the Internet recommends that if your cat has a wound, which you believe to be infected, you can sprinkle charcoal directly onto the area. This will aid in the healing process almost immediately! This remedy will help to clear up most any infection your animal may have.

Through his several years of health ministry in India, Bill saw many heart-rending cases. One case involved a man's ox. Oxen are very valuable to farmers in that culture, and are prized more than a tractor would be in America. The people are dependant on them for their livelihood.

This ox had had a boil on its rump, which had abscessed and become infected. The ox was so sensitive to the pain that they had to tie it down to work on it. Bill cleaned the area around the open wound, and then applied hot and cold packs to the area. He then packed the open ulcer with pulverized charcoal. The next day when the farmer brought the ox to be treated, it was quite willing to lay down on its own, to be treated. Eventually the wound completely healed.

With their permission, I'm sharing these stories from the Doctors Thrash:

> "A lady veterinarian told us of a horse which injured his foot by kicking the corral fence. The wound was swollen, infected, and painful. We gave the vet some charcoal to sprinkle into the wound. She was very pleased with the prompt improvement. Her antibiotics had not helped.
>
> "On the island of Okinawa, the habu, a well-known venomous reptile, is responsible for many deaths and disabilities every year. Just before our visit there, a pet kid, belonging to some friends of ours, had been bitten by a large habu. When they found it, its head and neck were extremely swollen, and the little goat, bleating inconsolably, could not stand.
>
> "Rather than abandoning it to its fate, some young men set about to try saving it. They decided on a remedy which they had used often on humans for wasp stings, spider bites, and poison ivy, but had never used for snakebites.
>
> "They dug a deep hole in the ground and filled it with a mixture of water, clay, and charcoal. The kid was laid in it, and the boys took turns holding the goat's nose above water. Although it seemed to get stronger, the boys soon tired, and since nightfall was approaching, they left the animal to fend for itself. Next morning they went back, expecting to find him dead. To their amazement, the kid had climbed out of the hole, and was grazing quietly nearby. Most of the swelling was gone. We can definitely say that charcoal and clay are useful for habu-bitten goats!

"Another incident involved a horse with a hopelessly infected eye. The horse was in a group of other horses being offered for sale, but because of its eye, it seemed destined for destruction. One family, intending only to buy a small pony from the owner, agreed that for an additional $25, they'd purchase the injured one also. They reasoned that the saddle was worth at least that much. Little hope was given for the poor animal.

"By the time the horse was delivered, the eye had become a sickening mass of oozing puss, with no evidence that an eye was even present in the socket. The family immediately went to work on the eye. Twice daily, charcoal poultices sprinkled with golden seal were applied to the eye. The poultices were held in place by a nylon stocking.

"Charcoal, flaxseed and apples were also blended together and fed to the filly. Within weeks the eye had cleared up, and she was as sound as the other horses. Less than a year later she produced a beautiful, healthy foal. As of this writing, there is no indication that she once had such an unsightly malady.

"A similar incident involved a Rhode Island Red hen which had been attacked by a neighbor's dog. She was placed in a cardboard box to await her fate. Although no wounds were visible, it did not, or could not, stand up, even when approached by strangers. The owners gladly gave the hapless bird to the first willing people who came along. This just happened to be the owners of the previously mentioned horse. She really looked fine … from the top.

"Upon arriving back at their own farm, the new owners took 'Hanna' out of the box to see why she refused to get up and scratch around with the other hens. They were not prepared for what met their eyes. Hidden under a writhing mass of maggots, lay the exposed labyrinth of her intestines. The smell was as foul as the sight. Their first reaction was to just put her out of her misery. But she sat there so patient, so trusting, so … well, full of faith, it seemed a shame to exercise less faith than a chicken, so out came the charcoal.

"Her bed was dusted heavily with the black powder, and she was fed liberal amounts of charcoal in her feed and water. Recovery, which included lots of time in the open air and sunshine, was rapid. Soon she was queen of the yard, playing mother to the six Banties sharing her domain. Charcoal had once again preserved health in the barnyard."[14]

Dr. Myatt frequently recommends charcoal for her two-legged patients at work but sometimes her four-legged pets at home also need the benefits of charcoal:

> "One of our cats somehow tore one of her claws out and the foot became badly infected. The vet recommended a treatment of antibiotics, but we decided to try a charcoal poultice first. My husband, Mark, was able to secure the poultice so that the cat could not remove it. By the third day the foot was back to normal."

Before her present job as program director for a massage therapy institute in California, Pamela made her home in northern Idaho, and kept horses. In Chapter 10 I shared some of Pamela's later experiences with charcoal, but it was this first encounter with the power of charcoal that made her a believer:

> "I first heard about charcoal in a seminar but didn't have much occasion to use it. We lived out in the country and kept horses. One day one of our mares fell and badly cut her knee. I was going away for ten days so I asked my husband if he would take care to it. However, when I returned nothing had been done, and the knee had swollen to the size of a grapefruit. It had healed on the outside, but it was obviously infected underneath. I did what I could, but it didn't help, so I called the vet. His verdict was that the horse was ruined.
>
> "It was then that I thought of using charcoal. I heated up some flax and water and made a jelly. Then I added charcoal powder. I plastered this jelly all over the knot. I covered that with an old bed sheet, and then ace-bandaged over it all.
>
> "After a week of changing poultices, the leg began to improve. The mare was a real pet. When I called, she would trot in from our 68-acre pasture. One day when she came up to me, I noticed the bandage had slipped down to her feet. The whole knot was gone! It had moved down to her ankle along with the poultice! I could not believe my eyes! After another couple of weeks the whole leg had healed completely. So much for being a 'ruined horse'. From then on I was sold on charcoal as a healing medicine."

If activated charcoal is not available at your local animal clinic or farmer's co-op, don't you think it's time you asked them to have it stocked? CR

---

[1] http://camell.atlas.co.uk/aahven.html

[2] George W. Poage III, Cody B. Scott, Matthew G. Bisson, F. Steve Hartmann, Activated charcoal attenuates bitterweed toxicosis in sheep, *Journal of Range Management*, Volume 53:73–78, January, 2000 http://uvalde.tamu.edu/jrm/jan00/poage.htm

[3] Struhsaker, TT, Cooney, DO, Siex, KS, Charcoal consumption by Zamzibar Red Colobus monkeys: Its function and its ecological and demographic consequences. *International Journal of Primatololgy*, 18: 61-72, 1997

[4] http://cliniciansbrief.com/cb/capsules/april03capsules.asp

[5] *Veternary Medicine*, September, 2003 http://www.aspca.org/site/DocServer/vettech_0703.pdf?docID=1044

[6] Jill A. Richardson, DVM, Christine Allen, CVT, ASPCA Animal Poison Control Center http:www.apcc.aspca.org

[7] Means, DVM, ASPCA *Animal Watch*, Summer 2001

[8] Dolder, Linda K, DVM, Metaldehyde Toxicosis, Toxicology Brief, *Veterinary Medicine*, March, 2003

[9] http://www.horseandfarmmagazine.com/vet1.html

[10] Cooney, David O, *Activated Charcoal in Medical Applications*, Marcel Dekker Inc., p.481, 1995

[11] Pass, MA, et al. Administration of Activated Charcoal for The Treatment of Lantana Poisoning of Sheep and Cattle, *Journal of Applied Toxicology,* 4[5]: 267-9, 1984

[12] Toxic Plants and Compounds http://www.pipevet.com/articles/toxicity.htm

[13] Cooney, p. 52

[14] Thrash, *Rx: Charcoal*, pp.53-55

Chapter 16 Environmentally Friendly

Our world, rugged but vulnerable, beautiful but fragile, shares its riches on condition. We do not own it. Nevertheless, we are its stewards. It is one thing to draft international environmental accords to protect our corporate heritage, it is quite another to get them acted upon. While governments and corporations waste precious time jockeying to protect their interests, the thoughtful attention that nature deserves can all begin right now with you and me… and charcoal.

Today, activated charcoal is used on a massive scale in both vapor-phase and liquid-phase purification processes. It is used widely in respirators, as well as in air-conditioning systems and in the cleanup of waste gases from industry. In the liquid-phase, its largest single application is the removal of organic contaminants from drinking water. Many water companies in Europe and America now filter all domestic supplies through granular activated carbon filters. Household water filters containing activated carbon are also in widespread use.

Other applications include decontamination of ground waters and control of automobile emissions. As a result of its commercial importance, charcoal has been the subject of an enormous number of research projects in both industrial and academic laboratories.[1]

The Water We Drink

"Water, water everywhere, and not a drop to drink. Water, water everywhere, and all the boards did shrink."[2] More and more, this famous ancient rhyme rings true for much of the world.

In her sobering article, *Drugging Our Water: We Flush It, Then We Drink It*,[3] Melissa Knopper reviews some startling realities. Birth control pills, estrogen replacement drugs, ibuprofen, bug spray, sunscreen, mouthwash and antibacterial soap, all pour into our drinking water supplies every single day. What's more, they could easily turn up in your next glass of tap water. In 2002, United States Geological Survey (USGS) scientists sampled 139 rivers and streams, finding the residues of hundreds of prescription and over-the-counter drugs and personal care products lingering in the nation's water supplies. In several cases, these contaminants were found in supposedly treated sewer water that ended up in the drinking water of another city downstream.

Some rural homeowners who use well water are at even greater risk. USGS researchers also turned up antibiotics in nearly half the streams that were sampled, raising other concerns about the nation's growing antibiotic resistance problem. "This study raised a bunch of red flags," says Dana Kolpin, lead author of the USGS study. "At these low concentrations, I think there are going to be long-term effects that may take several generations to show up."

William Owens, a toxicologist who researched estrogen patches for Proctor & Gamble, says that because the kidneys tack on an extra sugar molecule before they are excreted, until recently, people thought the estrogens in birth control pills were rendered inactive by the body. But now, scientists have learned bacteria in sewage treatment plants can 'chew' off that sugar molecule.

A British researcher, John Sumpter, studied fish living near a London wastewater treatment plant. He found male fish that were producing eggs. After finding the compound estradiol in the fish tissue, he concluded that estrogens from birth control pills were part of the problem.

Another active area of research and debate is antibiotic resistance. The Union of Concerned Scientists says farmers account for seventy percent of the antibiotics used in the U.S. Large factory farms use antibiotics to prevent confined, crowded livestock from getting sick. But that practice is creating 'superbugs', such as virulent strains of salmonella that are difficult to treat and can be deadly to humans. Dr. John Balbus of Environmental Defense, says that these superbugs may spread to consumers through contaminated meat or wells.

Meanwhile, thousands of pounds of triclosan—the active ingredient in antibacterial soaps, deodorants and household cleaners—are also going down the drain into our waterways. They may also be actively culturing antibacterial resistance.

The days of drinking pure water from a cool, sparkling, mountain stream are fast disappearing. The U.S. Environmental Protection Agency estimates ninety percent of the world's fresh water is unfit to drink without first being treated. As noted earlier, the WHO has calculated waterborne gastrointestinal infections alone, kill more than 50,000 people around the world every single day.

So what can be done? Polluted water can be made fit to drink in several ways including boiling, treating with chemicals, filtering, or irradiating it. The method(s) you choose will depend on what resources you have at hand, including cash, how often you need to treat water, and the amount of water you want. Each of these methods has its limitations. Boiling, treating with chemicals and irradiating may not kill, neutralize or remove all organisms and toxins, and they do nothing for the visible pollutants and odors left behind. Many filtering systems will restore clarity to the water and remove a large percentage of pathogens, but leave others.

Unfortunately, the answers to global pollution are no longer simple ones. However, while different stakeholders maneuver for tighter environmental protection legislation, one immediate remedy is accessible to all. Charcoal filtering systems that will remove the final traces of those pharmaceutical byproducts from our drinking water are readily available. In combination with certain iodizing chemicals, activated charcoal filters turn otherwise filthy, lethal water back into the closest thing to pure clear mountain water you will get. Depending on one's needs, they can be pricey, but in the short as well as the long term, they may prove to be our least expensive option. While we all wait for governments to implement environmental reforms, there is a solution near to your home.

If you have not discovered it already, you will find there are a host of charcoal filters of every size and shape, for every conceivable water purification requirement. Thankfully, water purification technology has come a long way from charred water barrels and hunks of charcoal in glass beakers. From domestic to recreational, from industrial to medical, from rural to civic, and from submarines to outer space, there is a charcoal water filter designed just for your needs. But, if you ever find yourself up a creek, with no fluorinated, chlorinated, charcoal-filtered city water, remember, a chunk of charcoal from your campfire in your pot of boiling water, will make it as safe to drink as it was for your great granddaddy.

"Water, water everywhere, and not a drop to drink". And so it would be for multiple millions of people, if not for charcoal.

The Air We Breathe

Never mind under the ocean or out in space, what about the air right around us? How can charcoal make breathing easier?

In Chapter 4 we reviewed charcoal's historical use for heating and cooking. Using lump charcoal does have advantages over briquettes. It is an all-natural, one hundred percent hardwood product, without the additives that are used to make briquettes. Natural charcoal heats faster than briquettes, so food can be cooked over natural charcoal as soon as it begins generating heat, usually within five to seven minutes after lighting it. Because there are no binders, lump charcoal can be lit with just a match and a piece of newspaper. This eliminates the need for lighter fluid, the burning of which can impart taste to food.

The fact that natural charcoal can be easily ignited allows users to start with a small amount and add to the pile, as more heat is needed. Natural charcoal can be smothered by closing off the air supply or putting it out with water, then reused at a later time. Finally, hardwood charcoal producers generally claim that natural hardwood charcoal retains heat longer than briquette charcoal, and that it is a more efficient fuel: one pound of hardwood charcoal produces heat roughly equivalent to two pounds of briquette charcoal.

These facts will probably not mean much to the average North American, unless they live on barbequed food. However, much of the world cooks with charcoal, and it is a good thing. It is a far cleaner fuel than plain wood, charcoal briquette, or coal, which means fewer pollutants, and cleaner air.

But, for the majority of us, as far as the air we breathe is concerned, the benefits of charcoal go far beyond what we barbeque with. Charcoal is working behind the scenes, cleaning up most of the air we pollute, not just outside around chemical and steel plants, but right inside our homes!

HEPA Filters

Today, just as we have an array of different charcoal water filters, so too, we have at least as many variations of charcoal air filters. HEPA (High Efficiency Particulate Air) filters, the Rolls Royce of charcoal filters, are the highest efficiency air filters available for the filtration of small particles. Defined by the Institute of Environmental Science, a certified HEPA filter must capture a minimum of 99.97% of contaminants at 0.3 microns in size. The first HEPA filters were developed in the 1940s by the US Atomic Energy Commission, to fulfill a top-secret need for an efficient, effective way to filter radioactive particulate contaminants. They were needed as part of the Manhattan Project, which developed the atomic bomb. HEPA filter technology was declassified after World War II and then manufactured for commercial and residential use.

HEPA air filters have been traditionally used in hospital operating and isolation rooms, in the pharmaceutical industry and in the manufacture of computer chips, as well as in other applications requiring 'absolute' filtration. Today HEPA air cleaners, vacuum cleaners and air filters are used in a wide variety of critical filtration applications in the nuclear, electronic, aerospace, pharmaceutical and medical fields. They are required by law to be used in all equipment for asbestos elimination.

Filters that meet the strict Military Standard 282 HEPA filtration efficiency test, are highly recommended for allergy and asthma sufferers. This efficiency rating means that the filtration systems are capable of removing those harmful particles, including dust, mold spores, dust mites, pet dander, and plant fibers that are the cause of many allergy symptoms. HEPA filters can even trap deadly Anthrax spores.

Another benefit of a HEPA air filtration system is that it can remove harmful V.O.C.s (Volatile Organic Compounds). These gases come from household chemicals and synthetic materials. HEPA filters are also advertised as helping to remove over one thousand identified indoor air pollutants, of which at least sixty are carcinogenic. These can build to dangerous levels in today's tighter, more energy efficient buildings.[4]

Odors

Activated charcoal is used in waste incinerators in a city near you. Many new cars use charcoal as the main ingredient in their air circulation filters. But this is nothing new. In 1854, Stenhouse described the successful application of carbon filters for removing vapors and gases in the ventilation of London sewers.[5]

You can make good use of that bit of history right in your home. If you don't want to buy a commercial product, just take a five-pound bag of charcoal, put half in a container with a perforated top, and place it in your fridge. Place the other half in your freezer for seven days. Odors will leave never to return!

I found this recommendation on the Internet. "When my friend put her house contents into storage, the woman in charge of the unit told her to put an opened five pound bag of charcoal near her furniture. She also told her to change it every six months. Apparently, the charcoal absorbs the odors and also the dampness in the air. Her furniture still smelled sweet after three years in storage. Hope this helps." Signed, Bob. Of course, the same can be done for closets, crawl spaces, and basements.

As for odors from domestic animals, just sprinkle charcoal on pet areas outside, include it in the cat litter box, or use it to soak up the telltale remains of a skunk visit. Perhaps you have had one take up residence under one of your buildings, but, long after you have removed the skunk, its calling card is still around. It is reported that a liberal sprinkling of charcoal powder will make quick work of any lingering odors.

Operating Rooms

Offensive odors come from all manner of sources, but no matter what brand, activated charcoal loves them all. However, some smells are 'invisible' to us. Operating rooms, whether medical or veterinarian, often have free-floating anesthetic gases lurking around. No one wants surgeons or their assistants to be dozing off. So, the bulk of these gases drifting around in the operating rooms are grabbed up by scavenging filter systems developed to remove them. As well, canisters containing activated charcoal are used as waste-gas disposal systems in lieu of other types of scavenging systems, especially when portability is an issue. For operating room personnel who are at special risk (e.g., pregnant), some institutions issue masks with activated charcoal filters.[6]

Laser Plume

Lasers in surgical settings are notorious for producing copious amounts of noxious smoke or plume as a by-product of vaporization. It has been proven that laser smoke contains both dead and live cellular material and viruses. Several hazardous contaminants and by-products have been identified. Because of the extreme temperatures used, vaporized particles from the destruction of the cell wall, become airborne. Some have been identified as intact cells, cell parts, blood cells, and viral DNA fragments. You can see the potential for infection.

Potentially toxic chemicals are also known by-products of laser smoke. Researchers have identified more than six hundred organic compounds in plume generated by vaporized tissue. For many of these compounds, there are harmful side effects including irritation to the eyes, nose, and respiratory tract; liver and kidney damage; carcinogenic cell changes; headaches; dizziness; drowsiness; stomach pains; vomiting and nausea; and rapid breathing. But, here too, charcoal is being used as a filtering system to minimize exposure to these potential health hazards.[7]

Industrial Waste

Those looking to get into environmental research and development will find that filter-grade charcoal is a fast expanding market. Calgon Carbon of Pittsburgh is one of the world's largest producers of activated charcoal. The company continues to show yearly gains in annual sales. One of the most important growth markets for filtered charcoal is with municipalities. Cities use filtered charcoal to remove chemicals, heavy metals, and other materials from the leachate from landfills. But there are larger problems.

Here, where I live in Nova Scotia, is home to the infamous Sydney Tar Ponds, the waste dump for the once thriving steel and coke industry. It is, for the moment, North America's largest toxic waste site. A cool two billion dollars have been allocated for its clean up. The project is touted as becoming the protocol for all future major toxic site clean-ups. But there are other sites, including Washington's Hanford nuclear waste site, New York's Hudson River, with other major waste sites-in-waiting, including Wisconsin's Mole Lake mine.

What I found disturbing, as I searched the various environmental groups' websites, was that virtually the only references to charcoal had to do with barbecue events. How is it, I wondered, that these bastions of environmental activism, make little to no mention on their sites of the wonders of the single greatest detoxifier in the world's history? Are they too so un-informed? The only allusion to charcoal's purifying nature was a brief mention by environmentalist John Muir in his book *Steep Trails.* "I found about a bucketful [of stagnant water] in a granite bowl, but it was full of leaves and beetles, making a sort of brown coffee that could be rendered available [drinkable] only by filtering it through sand and charcoal. This I resolved to do in case the night came on before I found better." I am not surprised when charcoal is overlooked by the general public, but it seems the environmental educators also need a seminar on the powerful virtues of charcoal.

Meanwhile, the Belgian steel producer, Sidmar, part of ARCELOR, the largest steel group in the world, won the Belgian Environmental Prize in 2002 for its substantial reduction in dioxin emissions using a new purification technology. This technology is based on the injection of activated charcoal into the released flue gases, causing the dioxins in the gases to be adsorbed.

Fuel Cells

With concern over fossil fuels, their costs economically and environmentally, industry has widened its exploration for safer fuel alternatives. Extruded, metal-impregnated, activated carbon has recently been introduced to the fuel cell industry. Fuel cells are being advertised as the fuel source for the upcoming new generation of motor vehicles. They produce power electro-chemically. Natural gas is one important feedstock for the production of hydrogen for these catalytic fuel processors. The sulphur that is often present in gas has detrimental effects on the catalysts. NORIT, another leader in charcoal production, developed a special activated carbon with a pore size distribution that efficiently adsorbs the common sulphur compounds.

Nuclear Power Plants

In nuclear power plants, the exhaust from the very tall stacks is sometimes radioactive. If unacceptable levels are reached, special fans automatically start, the normal ventilation system is shutdown, and the exhaust is routed through charcoal filters, and special high efficiency particulate, before being released.

Warfare

As mentioned earlier, research is ongoing to see if charcoal can, on a larger scale, contain chemical warfare. Charcoal-impregnated cloth is presently used in military uniforms worn by soldiers deployed in certain areas at high-risk for chemical weapons.

Buildings

As we have seen, global technology has gendered global problems. This is also true for the building industry. To conserve energy, we have designed airtight thermally efficient houses. But, to keep building costs down, and to better manage raw materials, more and more of the building materials are laminated or glued. It is ironical that three thousand years ago Egyptians used charcoal for embalming the dead, while today new houses are being 'embalmed' for the living. These glues, plastics, and insulators, which may include formaldehyde (also used to pickle organs for laboratories), can 'ooze' for years as they slowly decay from UV radiation. Sick air, sick water, sick people, sick animals, and now we have sick houses, all being treated with charcoal. A month rarely goes by, without another house, school or office being condemned because some mysterious mold or agent is making the residents sick. But, once again there is a natural remedy.

One Japanese company has started using charcoal obtained from bamboo to combat 'sick building syndrome'. Mayumi Yamada says her husband suffered from headaches, nausea and frequent illnesses due to the formaldehyde and other chemicals in modern building materials. But since she got the walls of their home treated with charcoal made from bamboo, she says he is now seldom sick. "My husband is prone to allergies and he complained that he couldn't sleep because his eyes were hurting. He has hardly complained at all since we had the walls treated," Mrs. Yamada said.

This method of coating walls with charcoal before they are plastered was the brainchild of Masanori Takao. He wanted to find ways to combat sick building syndrome after hearing so many homeowners complain about it in recent years. "I looked at various materials, but none of the ready-made ones would do the job, so I decided to make something myself. As such, I focused my attention on the properties of bamboo," he said. After mixing with a gelling agent, the bamboo charcoal is applied to surfaces.

Tests showed that the charcoal is able to adsorb toxic substances such as formaldehyde, which are present in walls. In the Yamada's living room, the coat of charcoal has reduced the level of formaldehyde by two-thirds, well within Japan's health safety standards. Although charcoal coating is about three times more expensive than plastic wallpapers, prices are expected to drop once demand for this product increases.[8]

Not only is the building industry using charcoal to treat sick houses, they have also discovered its insulation qualities. Here we have a perfect case of "killing two birds with one stone", insulate and detoxify your home with charcoal insulation!

Agriculture

People, animals, air, water, houses, what can charcoal do for the earth itself? All those pollutants, thousands of them, that aren't filtered from the air and water, end up in the earth.

The North Carolina Cooperative Extension Service in drafting their *Using Activated Charcoal to Inactivate Agricultural Chemical Spills*, states: "Activated charcoal is the universal adsorbing material for most pesticides."

This next bit of information is part of a fact sheet put together by the University of Florida:

> Sometimes it becomes necessary to stop the activity of an applied herbicide, perhaps because of an accidental spill, perhaps because of a weed-control and grass-seeding combination. Activated charcoal adsorbs one hundred to two hundred times its own weight and comes in handy for binding, thereby deactivating some herbicides. Turf areas that have been treated with pre-emergence herbicides can be reseeded earlier than normal by treating with activated charcoal.

Activated charcoal will reduce the level of most organic pesticides in the soil, but is considered ineffective for inorganic pesticides and for water-soluble organic pesticides. It is a good idea to keep a bag or two of activated charcoal in stock at all times so it can be applied almost immediately after an accidental spill or application. If the active material has not been diluted with water at the time of the spill, apply the charcoal dry. If it has been diluted with water, apply the activated charcoal in a slurry.

The charcoal must be incorporated into the contaminated soil, preferably to a depth of six inches. With severe spills, some of the contaminated soil may need to be removed prior to the activated charcoal application. It is easier to apply activated charcoal as a water slurry, so this is the best way to go when possible. The final spray mixture should contain one to two pounds of charcoal per gallon of water, and there should be enough water to begin moderate agitation until a uniform mixture is attained. Maintain moderate agitation while spraying.

For reducing the effects from spills of organic pesticides, some petroleum products and hydraulic fluids, use one hundred pounds of activated charcoal for every pound of active material spilled, but no less than two pounds per 150 square feet (600 pounds per acre) of contaminated area.

For turf areas that have been treated with pre-emergence herbicides, apply charcoal slurry at a rate of one pound per gallon of water for each 150 square feet. Wash the grass free of any heavy charcoal deposits, and, ideally, rake the charcoal into the soil. The area can be reseeded twenty-four hours after treatment.

To avoid the mess of a fine-powdered charcoal, look for a granulated product that dissolves easily. It can be spread by a walk-behind spreader without dust or irritation.[9]

Plant Poisons

In the last chapter, we mentioned the problems that horses have due to jugalone-contaminated sawdust. Jugalone is a natural hormone produced by black walnut trees, and is toxic to the roots of plants that encroach on the walnut's space. Moreover, when walnut trees are cut down, the decaying roots still produce the poison, causing a build up of jugalone in the soil. Charcoal can be spread around the area as a thick slurry and washed into the soil, or it can be worked in.

Irrigation

Just as the ecosystems of nature are interconnected, so too there are related issues around environmental stewardship. In California, alfalfa production uses more water than any other crop—almost twenty percent annually of the state's agricultural water use. Because of the use of irrigation systems, there is an increasing concern that unacceptable amounts of chemical fertilizers, herbicides, and insecticides will end up contaminating the already compromised ground water supplies. Recommendations are now in place urging the use of activated charcoal and other filtering agents, in conjunction with different containment strategies, to hold and clean surface-water run-off before water leaves a ranch into ditches, irrigation canals, ponds and rivers.[10]

Plant Food

While charcoal helps to clean the soil of pollutants, it also acts as a soil conditioner. It is used as a top dressing for gardens, bowling greens and lawns. Charcoal also acts as a substitute for lime in soil additives because of the high potash content, and it can be a little cheaper than lime. It is used for potting and bedding compounds as a soil and mulch sweetener, and as a fertilizer and insecticide for roses. Some orchids seem to love it. One study showed that adding charcoal to the rooting medium of peas produced a marked increase in the weight of the pea plants and in nitrogen fixation by the plants as compared to controls.[11] It is suggested that the benefits derived from charcoal are due to its adsorption of toxic metabolites that are often released by plant tissues, especially when the tissues are damaged.[12]

Here are some planting tips using charcoal chips. Start with a plastic liner in a tray. Add half an inch to an inch of gravel in the bottom for drainage. Next, sprinkle enough charcoal chips to cover the gravel layer. Charcoal will help keep bacteria at bay. Top this with potting soil and add your plants.

Insect Killer

Dating back to 1947, several studies have been conducted showing the benefits from activated charcoals in protecting seeds, seedlings, and crops from some organic pesticides and from the effects of herbicides applied to the soil to inhibit weed growth.

One study demonstrated that, as an insecticide, powdered charcoal is a more potent deterrent to the Tribolium castenum beetle, than are powdered clays. Commonly known as the Red and Confused flour beetles, these pests attack stored grain products such as flour, cereals, meal, crackers, beans, spices, pasta, cake mix, dried pet food, dried flowers, chocolate, nuts, seeds, and even dried museum specimens.

In fact, these beetles are considered two of the most damaging pests of stored products in the home and in grocery stores. It is speculated that the superior bleaching and desiccating properties of powdered charcoals account for its success in killing these pests.[13]

Odds and Ends

Drop a couple of pieces of charcoal into the bottom of your toolbox. The charcoal will absorb moisture, keeping the tools dry and rust free.[14]

Moistened charcoal granules have been used to scrub phosphine gas found inside large grain storages facilities. Phosphine gas, used to fumigate for rodents and insects, is both extremely toxic and explosive.

Charcoal is recommended for artists as the preferred clean-up agent for toxic spills of paints, resins, lacquers, glues, etc.

Photo finishing and chemical labs are often plagued with very toxic air-borne chemicals. Besides air circulating systems, some labs hang blankets woven with charcoal to the walls.

While charcoal promotes the growth of certain bacteria, it dramatically inhibits others. According to a joint research team from Japan's University of Osaka Prefecture and Okayama University, refined activated charcoal has been found to be able to absorb the O157:H7 strain of E. coli bacteria and its toxin that causes food poisoning. It is hoped that the research will contribute to the development of an alternative treatment in fighting bacteria without the use of antibiotic drugs.[15]

Early in the book I mentioned one Chinese company that sang the praises of charcoal. They cited a long and varied list of household benefits. In the kitchen, charcoal-filtered water makes cooked rice, stews and vegetables softer, fresher, and tastier. It removes strong odors when salted on fish or meat, and preserves them longer, without in any way affecting their natural flavors. It is a household prevention against mold. It acts as an air freshener and food preservative in the fridge, as well as for the gold fish jar. In blankets and pillows, it makes sleep more restful, and in shoes it eliminates unpleasant foot odors. As you can see from this short list of household aides, the Oriental culture lives up to its reputation of being very innovative. But charcoal's versatility stretches far beyond any one culture, as you will quickly see when you search the Internet.

In conclusion, there are over six thousand known toxic chemicals that pollute our environment. They enter our bodies through the air we breathe, the water we drink, and the food we eat. For instance, up to the time of harvest, apples and grapes can have over twenty toxic chemicals used on them. But we do not have to be hostages to these poisons. Charcoal can work for you just as it does for steel manufacturers and hospitals.

Every home should have charcoal on hand. It can be used as an antidote for poisoning and as a cleansing agent for our families, our animals and our homes. It can also serve to benefit the water, the air, and the land we take so much for granted. In short, it can help us sweeten and refresh this awesome world our Creator has so graciously given us. CR

---

[1] http://www.personal.rdg.ac.uk/~scsharip/Charcoal.htm

[2] Coleridge, Samuel Taylor, The Rime of the Ancient Mariner, 1858

[3] Knoppe, Melissa, Drugging Our Water: We Flush It, Then We Drink It, *E/The Environmental Magazine,* Volume XIV, Number 1 http://www.emagazine.com

[4] http://www.vacuum-cleaners-ratings.net/hepa-filters.htm

[5] Carbon Materials Research Group, Center for Applied Energy Research, University of Kentucky, Historical Production and Use of Carbon Materials, December 5, 2003

[6] Members of the ACVA ad hoc Committee on Waste Anesthetic Gas pollution and Its Control, American College of Veterinary Anesthesiologists (ACVA)

[7] Andersen, Karen, Safe use of lasers in the operating room: what perioperative nurses should know, *AORN Journal*, Jan, 2004

[8] Fu, Jinhe, PhD, Program Officer International Network for Bamboo and Rattan (INBAR) http://www.channelnewsasia.com/stories/eastasia/view/47358/1/.html

[9] McCarty LB, Fact Sheet ENH-88, Adapted from "Activated Charcoal for Pesticide Deactivation", Cooperative Extension Service, University of Florida

[10] Cline, Harry, *Western Farm Press,* Jan 24, 2004

[11] Vantis, JT, and Bond, G, The effect of charcoal on the growth of leguminous plants in sand culture, *Annals of Applied Biology*, 37, 159, 1950.

[12] Cooney, David O, *Activated Charcoal in Medical Applications*, Marcel Dekker Inc., p.559, 1995

[13] Majumder, SK, Narasinhan, KS, and Subrahmanyan, V. Insecticidal effects of activated charcoal and clays, *Nature*, 184, p 1165, 1959

[14] Faulk, Tracy, Toolbox rust prevention, *Family Handyman*, June, 2003

[15] Charcoal can absorb E. coli, *Food & Drink Weekly*, Oct 9, 2000

Chapter 17 On the Lighter Side

In looking at health, total health, vital health, I do not want to leave the impression that charcoal is a cure-all. While charcoal can be used as a stand-alone treatment, it does not stand apart from a holistic program of preventive medicine. There are no magic bullets, not even black ones. Having said that, there are other simple natural remedies, many of them. In the previous chapters there have been several references to other simple cures. In this chapter we will take a little closer look at some of them.

I'm sure that many of you have a recollection of a simple remedy given to you when you were small. While my mother practiced her healing "art" on me when I was young, my first real awareness of the power of simple natural treatments came as an adult.

I was on my around-the-world hike. I was reveling in the warm exotic climate along the Mexican Pacific coast. After two months of backpacking across the Baja desert I was still feeling great, until I finally came to a tourist area.

Within a couple days my passport got stolen, and the next morning my bowels felt strange. By that afternoon I knew my fears were real. I was another casualty of the "Tourista Blues" (a.k.a. severe diarrhea). You have to have suffered through a case of tropical dysentery to appreciate how quickly the feeling of being in paradise can dribble away. "Much discomfort in the abdomen" does not at all describe the feeling.

My few acquaintances suddenly vanished. I knew I could not just lie suffering in my pup tent. I had heard that there was a small waterfall that spilled onto the beach about a half mile down the shore, and it seemed refreshing to my miserable state of mind. I slowly made my way along the sandy beach, resting every so often, until finally I came around an outcropping and there it was.

The clear cool water gushed over the bank, and spilled the twenty feet onto a large flat upright rock embedded in the sandy gravel below. It was the perfect bench. I stripped and sat down under the refreshing spray. Instantly I felt better and would have stayed longer except some embarrassed women began to line up for their turn to wash clothes.

No doubt it was the local tap water that infected me, but it was also water that helped turn off the tap on my tourista blues. No, I did not know about charcoal then. Otherwise, I certainly would have used it, but I realized the healing power of water in a very dramatic way. Water therapy, or hydrotherapy, though as ancient as illness itself, is once again coming into its own.

Drugs

Why should someone even consider old-fashioned water treatments as a first line of treatment for disease instead of the convenience of drugs? There are numerous reasons, but the very nature of most drugs should preclude them as a first choice.

Dr. Lisa Landymore-Lim, (PhD Chemistry) has worked for the National Institute for Medical Research, London, England, and the Dunn Nutrition Unit, Cambridge, England. She says most of today's pharmaceutical preparations, because of their harmful effects, may be labeled "poisonous". Drugs are not only poisonous in the amount given, but in their very chemistry. It should also be explained that there is absolutely no energy in any drug apart from a few odd calories from sweeteners. A drug creates the allusion of giving energy by prodding some organ to give up some of its own vitality to produce some desired effect. This false sense of energy or healing is at the expense of the deprived organ.

There is a common fallacy that drugs cure disease. Drugs do not cure disease. Drugs work by manipulating some sequence of biochemistry in already existing body mechanisms. However, this often proves to be too narrow in scope, so another drug is added and another, along with their all too common toxic side effects. For example: A child develops a fever which alarms the parent. A drug is given to inhibit the fever mechanism and the temperature drops to normal. But, the drug has done nothing to cure the problem that prompted the body's defense mechanisms to fire up in the first place. Instead, all too often, the cause of the original disease symptoms is expelled from one organ simply to take up residence in the weakened organ and produces a different set of symptoms. Furthermore, the diseased body must now work to rid itself of the poisonous drug.

There is some suggestion this is one mechanism for Type I juvenile diabetes. A child is given a vaccine, or is prescribed a course of antibiotics, to quell some viral infection or flu-like symptoms. The difficulty seems to be overcome, but, at some later period, from a week to a month or even a year, the child suddenly presents with aggravated signs of pancreatic failure. The drugs have chased the disease from one area in the body to the pancreas where it destroys the beta cells that produce insulin. Apart from an aggressive recovery program at the very outset of symptoms, including a very strict vegetarian diet and hydrotherapy treatments to the pancreas, the pancreas will not recover, and the child will be consigned to a lifelong connection to needles and insulin pumps. It need not have been.

In her 1994 book, *Poisonous Prescriptions*, Dr. Landymore-Lim reports on investigations which found that diabetes might in fact be a major side effect of antibiotics and other common pharmaceuticals. The book provides evidence from studies and hospital records.

As I write, one pharmaceutical hails its newest answer to diabetes, the saliva from the Gila monster—the brightly colored venomous lizard of the southwestern U.S. On the other hand, natural remedies, such as water treatments, do not add any more poisons into an already distressed body. Rather, they supplement the body's own broad-based defense mechanisms.

Hydrotherapy

Dr. John Harvey Kellogg is best known today for his corn flakes invention, but few know he was also the director of the world famous Battle Creek Sanitarium in Michigan, from 1875 to 1942. The rich, the famous, and the common people from around the world came to be treated for all kinds of ailments with any one of his many different water treatments. Baths, showers, sprays, fomentations, whirlpools, saunas, massage, compresses, packs, and other water treatments, encompassed his hands-on therapy. While some of these treatments were quite elaborate, most were simple enough to be duplicated in the home. This made them both practical and affordable.

Erna is familiar with charcoal as a remedy, and she also understands how to stimulate the body's own healing mechanisms with simple water. She shares a few more of her experiences.

In Chapter 9, Erna related her daughter's experience with codeine poisoning, which shut down her bowels. After relieving her distress with charcoal taken orally, there still was the problem of elimination. The nurses had tried an enema, and told her to just eat and wait for her bowels to start again. However, it had been five days, she was losing weight, and Erna was worried:

> "In order to get her digestive system functioning again, I tried doing hot and cold treatments on her abdomen. I took some face cloths and soaked some in hot tap water and one in ice water. I applied the hot clothes for four minutes and the cold cloth for one minute. I repeated the sequence about seven times and within an hour and a half her bowels were back working.
>
> "The pain from her broken leg was very bad—especially after we quit the codeine because of the effects it was having on her digestive system. In order to give her relief, I did the same hot and cold treatments on her injury site. I would repeat them whenever the pain returned—sometimes every few hours. It worked wonders for her.

"I crushed a vertebra in a car accident. Because it was a stable fracture, they sent me home with painkillers and told me to lie down until I felt that I could get up. I was in a lot of pain, but the best relief I would get was when someone would give me hot and cold treatments on my back. Usually I was home alone with the kids, so it was up to them to do the treatments. They'd fill a hot water bottle with very hot water for the 'hot' part, and used a gel pack from the freezer, for the 'cold' part. After a couple of rounds of alternating hot and cold, I'd be comfortable again.

"Most people predicted that I'd have lasting pain from that injury, or be incapacitated for the rest of my life in some way. But, it healed up beautifully and, except for when I'm sewing, I never even have a twinge. I really feel it was because of the hot and cold treatments. They increased the circulation to the injury and helped it heal more quickly and more completely.

"One spring our son Justin had an allergic reaction to some unknown. One side of his face swelled up, and felt jelly-like. He also broke out in an itchy rash. When I took him to the doctor, I was given a prescription for an antihistamine and a cortisone cream to rub on the rash. I paid a painful amount for the medication, and then went home to give it a try. After several days the rash was gone but the swelling was at least as bad as ever.

"Before taking him back to the doctor, I decided to try a flax seed poultice. I ground up flax seed in the blender, and mixed it with water to form a paste. I smeared the paste liberally on his swollen face. It was tricky to keep it in place, but within a couple of hours the swelling had gone down considerably. I put another batch on, and I believe he slept with it all night. In the morning his face was back to normal."

Erna mentioned it was difficult to keep the blended flax paste in place. Heating the paste and letting it cool into a gel will help. After applying the gel, place some saran wrap over it, gently pressing it to mold with the face. This can then be held in place with a loose ace bandage or knitted cap.

This next story has similarities to other experiences, but notice how the treatment with charcoal is combined with water. The Boykins shared this experience that happened while they were working in Central America. "A man with a badly infected leg traveled some distance to a hospital for treatment. The doctors' only solution was to amputate. The man decided to go home to die. But, before he left, he talked with a nurse in his church, who then related his story to Nellie. She directed the nurse to pound up charcoal into a powder, make a poultice, and apply it over the infected area. They applied a fresh poultice every day, and also applied mild hot and cold wet packs to the whole leg. They continued this routine daily for three months, and gradually the ulcer healed over completely."

Hot Foot Bath

Perhaps the simplest of the many water treatments is the hot foot bath (HFB). Using a basin big enough, immerse the patient's feet in hot water (38° to 43°C or 100° to 110°F) to above the ankles. You may also add some mustard leaves to the water. This will help to relieve a headache or any general congestion of the head. As blood is drawn to the feet away from the head, the congestion that often sparks a headache is dissipated.

In addition to the HFB, if the patient is wrapped in a blanket that is draped over the basin, with a cold cloth applied to the head, and with additions of more hot water to the HFB, this treatment can be used to produce an artificial fever. Why would one want to create an artificial fever? This may be an opportune place to explain a little physiology.

Physiology

We are so designed that when the body is invaded by some foreign pathogen, it responds by producing antibodies. These antibodies, in turn, activate the body's other defense mechanisms. Body temperature will climb, prompting the body to begin producing more white blood cells, which in turn attack the pathogens. As we heat up from the fever, our breathing increases. This increases the oxygen delivery to the entire body. As we breathe harder, our heart pumps faster, speeding up the circulation of oxygen-rich-blood, laden with white blood cells. These are just a few of the automatic, built-in defense mechanisms designed to help us combat disease.

Unfortunately, most folks are uninformed about what happens when a fever sets in. They often panic, supposing the fever is the disease, when in fact, the fever is a sign that the body is doing exactly what it was designed to do—fight disease. Our body functions, including temperature, cycle regularly between higher and lower levels, and, if our immune system is strong, we hardly notice. But, when, because of a poor lifestyle or other cause, our immune system is compromised, our body's defenses can quickly be overcome by the invading pathogens, and we get "sick". Our defense mechanisms may overreact, and the fever runs out of control, doing more harm than good. Or, they may react sluggishly allowing the disease to slowly progress. Often, the body's immune system and the disease end in a stalemate. For one reason or another, the body's defenses are never fully activated, and the disease percolates and becomes chronic.

Under these conditions, by producing an "artificial" fever, we can stimulate the body to react more aggressively. Using different treatments such as hot baths, or hot wet packs, or different combinations of hot and cold contrasts, the body can be artificially stimulated to increase its immune system and the delivery of its own "antibiotics".

HFB continued…

Before going to the next story let me finish outlining the HFB used to produce a general artificial fever. Give the patient a glass of hot water to drink as you begin the treatment. Keep the HFB as hot as the patient can tolerate by raising the blanket and adding more hot water as the treatment proceeds. Have them put their feet to one side of the basin and add the hot water as you agitate the water with your hand so as not to burn them. Replace the blanket.

When the patient begins to sweat, apply a cold cloth to the head and change it often. Continue the treatment for fifteen to twenty minutes after sweating begins—five to ten for younger patients. When completed, throw back the blanket, and, before removing the feet completely from the HFB, elevate them and pour cold water over one foot then the other, and then set them on a dry towel. Have the patient stand and give them a sponge bath and pat them dry, or they can take a tepid shower and dry off. Then let them lie down in bed and cover them with a blanket, allowing them to slowly cool off. Often the patient will fall asleep.

If the patient is fighting a systemic infection, the HFB will raise the autoimmune system to aid in combating the disease. If they are experiencing a low-grade fever that lingers on and on, the HFB will often stimulate the body such that the fever is broken.

Caution: it is not recommended to give a hot foot bath to patients with poor circulation to their feet (as with diabetics, patients with arteriosclerosis or with Buerger's disease) or those with loss of feeling in their feet or legs, or are unconscious.

Pastor Gary holds a Master's degree in Public Health and believes healing and teaching should go hand in hand. Here are two of his many stories. The first is a variation of the HFB:

> "I was visiting a woman with an infection in her right arm. The arm had swollen to half again as big as the left arm. There were no lesions or abrasions. Four times a year she would go in for antibiotic shots for the infection.
>
> "When she mentioned infection to me, I asked if she had heard of hot and cold applications. She hadn't, she said, but she was willing to try. Since she lived on a farm, I was able to find a large laundry tub. We filled it with hot

water, and immersed her arm in it, up to her shoulder. We also cleaned out an old ten-gallon milk container from the barn, and filled it with cold water.

"She sat on a stool and immersed her arm, first for five minutes in the hot water, and then for one minute in the cold. We did three changes. When we were done with the treatment I had a prayer with her. She called the next day to say that it was the first time she had slept well in weeks. When I met her the following year, she threw her arm up over her head, and said there had been no infection since the treatment. Later, I was told that sometime before I met her, she had had a mastectomy, and many of her lymph nodes had been removed.

"I went to visit one family that had been attending my meetings, but the husband would not come out of his room. When I asked what the problem was, the wife said that he was embarrassed to be seen. It seems he had overdosed on niacin, and his legs were inflamed and terribly itchy. His legs and feet were so sensitive that the only way he could get relief was to lie on a table with his feet hanging off the edge. I directed them to use hot and cold treatments [as above] and he very quickly found relief."

Ice Baths

By and large, most people have found that heat, in one form or another, is the treatment of choice when dealing with these kinds of infirmities, but not always. Sometimes, when patients fail to respond to one form of treatment, they do improve with another.

John was forty-four years old and in good health when he fell eighteen feet, crushing his heel. "The pain was extraordinary. My foot ballooned and the skin got red and blistered. The pain was so intense I was not able to sleep.

It was wintertime, so I had Emma Jo fill a plastic tub with snow, and bring it inside. We topped it up with cold water, and I draped my leg over the bed, and into the bucket. For the first time in days, I was able to sleep. I slept that way for a week. During the day I would get up and stretch."

Burns

Burns, in general, are another injury that respond very well to cold or icy water, or a cold-water spray. I have listened to some amazing accounts of severely scalded children being immersed in cold water, who were quickly relieved of all pain and healed without any scars at all. To subject someone who is critically burned to cold or icy water seems to me to be cruel. However there is sufficient research to show that if a burn victim can immediately have the burned areas immersed in cold water, there is significant reversal of tissue damage, and inflammation and swelling are reduced, with little or no blistering or scarring.[1] The general advice seems to be the use of water temperatures between 50° to 58°F (10° to 15°C), or cold tap water for ten to thirty minutes.[2]

Heating Compresses

Phil has been the director of Silver Hills Guest House for many years, and has seen patients with almost every type and stripe of illness come to learn from their lifestyle program. Along with the holistic approach, he also employs the use of natural remedies, including hydrotherapy. When I asked him what his favorite treatment was, he did not hesitate. "The heating compress is by far my favorite. It is the easiest to apply, it stimulates the immune system, and soothes away stress."

The heating compress may be applied to the neck for a sore throat, to the chest for coughs and bronchitis, to the abdomen for constipation or abdominal pain, and to any part of the body for pain or infection. By using a knit cap to hold the compress in place on the upper face, it can even be applied to the sinuses.

Before going to bed, heating compresses can be applied around the abdomen and back for sedation, chronic active hepatitis, backache, ulcerative colitis, chronic indigestion, and chronic stomach problems. They can be used on the feet for a cold.

There is a seeming contradiction to the term 'heating compress.' The compress is actually applied cold, but within about five minutes, it heats up from the reaction of the body against the cold. The heat produced is captured by a wool wrapper, and used to increase the temperature of the tissues. The mechanism is similar to a diver's wetsuit.

A short treatment produces a tonic effect. A long treatment is used as a sedative, relaxing the muscles and associated blood vessels. Heating compresses are effective in relieving pain, sore throat, cough, boils, cellulites, abdominal distress, whooping cough, croup, pneumonia, and constipation.[3]

The following are some examples of how compresses may be applied:

Throat

You will need one or two thicknesses of cotton or linen. It should be wide and long enough to comfortably wrap around the affected area—say, 2 x 14 - 20 in. (5 x 35—50 cm) for the throat. Wet the cotton cloth in as cold water as possible. Quickly wring it dry and wrap it around the neck. Over the wet cotton cloth, place a piece of plastic large enough to cover the wet cotton piece by at least one half inch on all sides. This in turn, is covered by a wool flannel cut to fit the area and large enough to cover all of the cotton and plastic. A piece 3 – 4 x 34 in. (8 – 10 x 85 cm) will wrap around the neck twice. A wool scarf or sports sock will suffice.

Abdomen

For a moist abdominal binder, use a strip of woolen acrylic material, such as wool flannel or men's suiting material made from a grade of wool. The material should not be as stiff as felt, but relatively thin and easily workable. I should measure 12 x 55 in. (30 x 135 cm). A cotton piece used under the woolen or acrylic piece should measure approximately 50 x 10 in. (127 x 25 cm). Short strips may be sewed together.

Chest

For a chest-heating compress, the cotton underpiece should measure approximately 9 x 100 in. (22 cm x 3 m), and the woolen or acrylic piece approximately 10 x 100 in. (25 cm x 3 m).

Begin by dipping the cotton or linen compress in cold water, and then wringing it dry enough that it doesn't drip. Then mold it over the area being treated. Cover it completely with plastic, fitting it snugly, but not so tight as to cause discomfort. Be certain there are no portions of the wet cotton exposed.

For treating the chest, a large plastic bag such as a dry cleaning bag or a trash bag may be used. Cut a hole in the bottom for the head, and in the sides for the arms. Wear it as a pullover sweater. Wrap and pin the wool flannel securely over the first two layers. Pin any loose fabric into darts so that the end product is smooth fitting.

It may be left on from half an hour to overnight, or between other treatments. When the compress is removed, you may want to rub the area under the compress with cold water or rubbing alcohol. A heating compress to the throat may be worn in the daytime by wearing a nice scarf instead of the wool flannel.

Feet

For a heating compress for the feet, use two pairs of socks. The inner pair should be moistened in cold water. Next, put ordinary bread bags over each wet sock to hold in the moisture. Then put on the second pair of socks, rolled on over the bread bags. Be sure that the feet warm up well, and then they will be a few degrees warmer than usual all night.

Wet Sheet Pack

Another variation of the heating compress is a full wet-sheet pack. These are used as a general tonic in chronic illness, for infections or fevers, emotional events such as anxiety, depression or stress, and generalized dermatitis. This treatment is not recommended for faintness, phlebitis, boils or open lesions on the skin.

Again, the mechanism is a reaction to the cold that produces a general increase in skin temperature. Drawing internally congested blood to the surface will often break a fever. The general feeling of warmth is relaxing and sedating.

There are a couple of ways to apply the wet-sheet pack, standing or lying down. The wet sheet to be used in this treatment should not be left too wet, as it tends to change its temperature too rapidly when a lot of water is left in it.

Standing

There should be no drafts in the room when applying the wet sheet, and the patient should be warm. Secure a cold compress around the forehead at the beginning of the treatment. Wring a bed sheet from cold water, between 60° and 70°F (16° to 21°C). Begin by having the patient hold the sheet under one arm, then carry the sheet around the back, over the opposite arm, across the abdomen, covering the first shoulder and arm, and tuck it in at the legs and neck. Over this, wrap the patient in one or two blankets. Place a towel around the neck to cover any exposed area. If the wrap is done next to a bed, it is an easy matter to help the patient to lie down. Then the bottom ends of the sheet and blankets can be quickly tucked under the feet.

Lying Down

First, place a protective plastic sheet on the bed. Next place a blanket on the bed in such a way that the far edge hangs longer than the near edge. After dipping a sheet in cold water, wring it out as dry as possible. Spread this over the blanket so that the top of the sheet is a little lower than the blanket. Have the patient lie down on the sheet with the shoulders several inches below the top of the sheet. Hold the arms up and wrap the near side of sheet over the body and nearest leg and tuck it in. Drop the arms and pull the far side forward over the arms and far leg. Tuck the sheet under the near side. Fold the sheet over the shoulders and across the neck. Repeat the process with the blanket. Double the bottom of the blanket under the feet.

Place a towel around the neck to exclude any air. Cover the patient with a second blanket and tuck in the edges. A hot water bottle at the feet will speed up the heating process. Sponge the patient's face at the beginning of

> the treatment and then continuously after the sweating reaction has begun. This may be left on from thirty minutes to three hours. Most patients fall asleep after they reach the heating stage—sometime after ten minutes.

I have outlined only two ways that I have applied this treatment, but there are variations. The things that make for a good treatment are, first, having the patient warm before the treatment begins. Giving them a warm bath or HFB and a drink of warm beverage will help raise their temperature. Second, being sure the wet sheet makes uniform contact with the skin. Third, avoiding any chilling of the skin by securing the edges of the blankets. Fourth, keeping the feet warm throughout the treatment. If these conditions are ensured, the warming up process should begin immediately.

Joel shared this last minute story from Nepal showing how a hot treatment can be used effectively to break a fever:

> "A few days before we were scheduled to leave Nepal, one of our neighbors, who worked for a phone company, got sick with a high fever. He asked us to treat him. It was during a politically volatile time, so, even though we had been warned by the U.S. Embassy not to venture out, we went after dark. We arrived safely at the man's home and gave him a HFB. We put a kettle on a hot plate under his chair to speed up the fever. But he couldn't stand the heat. So we moved him into the bedroom, and his daughters helped me wrap him in a wet sheet pack. Then we wrapped him in blankets, and gave him hot tea to help spike the fever. He had trouble coping with the heat for the first few minutes, but it wasn't long before he fell asleep. I had a short prayer with the family, asking God to bless the treatment, and then we left. The next morning he came by to say, 'Your God healed me!' It was a nice farewell."

Fomentations

The application of hot fomentations is a simple treatment. It involves using three or more folded bath towels and used as fomentation packs. The towels are dampened and preferably steamed in a large kettle with a rack in the bottom. If this is not feasible, you can wring them out in very hot water or, depending on thickness, put them in a microwave for three to five minutes. These hot packs can be applied to any part of the body. When applying fomentations, you will want to be sure that the room and your patient are warm.

First place dry towels, from one to four layers, over the area to be treated. Quickly place the hot pack on the dry towel, and cover it with another dry towel. After three to six minutes, remove the pack and rub down the area for thirty seconds with a cloth dipped in ice water and wrung out. Replace the dry towel over the area.

Repeat the above steps at least two more times. Always end with the cold rub. Try to stay with the same time sequence throughout the treatment—i.e. four minutes hot followed by thirty seconds cold.

The first couple times you do this treatment, you will likely feel awkward. But you will soon see how simple the procedure really is. With intelligence, carefulness, and attention you will soon feel competent. The packs may be too hot and then you will need an extra dry layer between the pack and the skin, or it may not be hot enough. Ask the patient if they can tolerate the packs being hotter. Show them how to raise the hot pack in case it gets too warm if you are out of the room.

Sometimes fomentations are given with a HFB. The patient may hang their legs off the bed into the footbath, or they may lie out on the bed with their knees flexed. Either way, you will want to protect the bed from the sweating with a blanket or plastic sheet underneath the patient.

If the HFB doesn't work, you can wrap the feet with a hot pack or put hot water bottles around the feet. Then cover the patient with a sheet or blanket. Once they begin to sweat, you will also want to keep the head cool with a cold folded washcloth to the forehead or neck. Remember to finish the treatment by pouring cold water over the feet.

Fomentations are good for congestion of the lungs or sinuses, for pain relief, or muscle spasms. Do not use the HFB for people with poor circulation or those who are paralyzed or unconscious. It would be easy to burn them. If the treatment has worked as it should, the patient will feel relaxed, and may want to sleep.

As you can see, these treatments are simple, both for the patient and for the caregiver. The simple household equipment is affordable and available in almost every home. These treatments adapt themselves easily to emergencies as well as to less threatening day-to-day mishaps, aches, infections and viruses. This is practical physics and practical chemistry that even children may use. Mixed with a simple prayer of faith they will be doubly blessed of heaven. These represent some of the simple treatments that sustained many of those pioneers who braved unknown adventures as they settled across the new world. These are the natural treatments missionaries continue to take with them to distant lands. These are some of the simple remedies that have spanned the millennia of time bringing relief to those suffering in body and mind.

As individuals, we need to become more acquainted with how the human organism works, so that we can better cooperate with our body's defenses rather than drugging them. In the Appendix, I have listed resources that describe different home remedies in more detail, especially hydrotherapy treatments. Even though these methods have, with the introduction of quick-fix drugs, largely gone out of fashion, they have not gone out of service. In fact, there is a renewed interest in these rational methods of treating disease among certain medical schools.

Fomentations and Charcoal

After immigrating to Ireland and settling into their new home, John and Sharon began to visit in the homes of their new acquaintances. While visiting with one family, John, a health educator, asked if the man had any health problems? John relates what followed:

> "'Do I? No one can help my severe stomach pain. Wouldn't be talking to you now if I hadn't just taken pain medicine. I'm on disability. Can't even go out and play with my kids.'
>
> "Pat was in his mid thirties. He had terrible pain in his stomach, and the doctors had almost turned him inside out trying to find out what the cause was. Specialists had not been able to identify the problem.
>
> "'I think I can help you. May I try?' I asked.
>
> "'How much do you charge?'
>
> "'Since I'm not a medical doctor and use only natural remedies, the price is free.'
>
> "'Well, I've got nothing to lose. Let's give it a try.'
>
> "I did a lifestyle analysis and found that he was violating almost every natural law of health. As I taught him and his family the basic principles of health, they began by improving their diet. After one week, his wife, who had been chronically constipated, was feeling much better. Over time, they followed each natural law. Having discussed diet, I next offered to give Pat hydrotherapy treatments.
>
> "'I'll try it if you think it'll do any good,' he assented.
>
> "'I'll need the entire family to help,' I requested.
>
> "I gave the fomentations. Their son held the clock and timed the alternating application of heat and cold. The daughter kept his footbath hot, and the little one put cold cloths on his forehead. We had a great time working together. After three treatments, his wife knew how to proceed without my instruction. Soon they did a better job than I did. Following each treatment, I suggested we put charcoal poultices on his stomach.
>
> "'But we can't buy charcoal powder here, only charcoal sticks,' Pat told me.
>
> "At my request, they purchased some charcoal sticks. The entire family watched as I put the charcoal in a bag, ran over it with the van, and then

pounded the charcoal with a sledgehammer. Finally, I poured the charcoal into a blender to pulverize it into powder. This was made into a poultice and placed over his abdomen. We always followed each treatment with a prayer asking God to add His blessing.

"After about five weeks of treatments, and an improved lifestyle, Pat was off all drugs, and could move around without pain. What was it that worked, the change in diet, the fomentations, the charcoal, the prayer, or a combination of all four? I don't know, but I do know that Pat got better."

So the God of heaven and earth works today. He has placed within our individual horizons remedies that will meet almost every emergency. Pioneer health reformer and lecturer Ellen White (Chapter 18) wrote in 1890, "If we neglect to do that which is within the reach of nearly every family, and ask the Lord to relieve pain when we are too indolent to make use of these remedies within our power, it is simply presumption."[4] Two such simple remedies are charcoal and water. But, there are many others.

On the Lighter Side

Back here in Cape Breton, Dan complained of sinus congestion that had been bothering him for days. Smiling to myself, I told him to lie back in a comfortable chair, place a wet cloth over his eyes, and hold a light bulb over his forehead. He gave me a skeptical look. "**Don't** forget to plug it in. **Don't** hold it so close you burn yourself and **don't** stick your feet in hot water," I told him. How long did the dry heat take to completely clear his sinuses? "Less than fifteen minutes!" This simple treatment has worked repeatedly for different ones.

It also works wonderfully for earaches and is without peer in giving relief for the burning and discomfort of the perineum in the postpartum period after childbirth.

Herbs

In the late 1800s, when allopathic doctors were prescribing all manner of poisonous concoctions, Ellen White pointed people back to the simple agencies of nature:

> "The Lord has given some simple herbs of the field that at times are beneficial; and if every family were educated in how to use these herbs in case of sickness, much suffering might be prevented, and no doctor need be called. These old-fashioned, simple herbs, used intelligently, would have recovered many sick who have died under drug medication."[5]

One favorite herb of Ellen White was red clover tops. It has long been used as a 'spring tonic' for purifying the blood. Herbalists list it as one of four home remedies which have helped in the treatment of cancer. It is a sedative and recommended for ulcers.

My first experience with the healing virtues of herbs goes back to my time in Guatemala. Doctor Graves and I had gone to be baptized up in the mountains in a large cattle trough filled with frigid mountain water—God's divine hydrotherapy. As we drove along on our return trip, I mentioned my little ailment that had been bothering me for over a year. It was nothing major, just a nagging little cough.

Shortly after mentioning my cough, Doctor Graves pulled over and directed me to some blue/purple flowers growing along the roadside. I scurried up the hillside and picked a bag full. We then resumed our trip back to the mission. Later he made me a large cup of tea made from the blue/purple leaves and had me drink it. After that he had me take a light box treatment. It consisted of a four-foot by four-foot cube box with four banks of light bulbs in each corner, with a hole in the top to stick one's head out of. It guarantees a good sweat. After that good sweat I took a cool dip in the river.

The next morning my cough was gone, and it has never come back. Many have asked me what the blue/purple flowers were. I'm sorry that I cannot remember. Whatever they were, it was clear to me, after more than a year, that the combination had worked like magic—light purple magic.

One herb I have used and know the name of is smartweed. Its botanical name is *Polygonum hydropiper*. It is a common garden weed with pointed leaves and small pink or white flowers, and goes by several different common names. It is recognized by its peppery flavor. It is reported to relieve pain.

I have personally used it for toothaches and recommended it to others. The leaves can be picked and used fresh from the garden, or dried and stored in airtight jars. The fresh leaves are easily chewed, and for an offending tooth, simply maneuver the pulpy mass next to the tooth. Often the pain from a toothache is significantly relieved within ten minutes. Smartweed can also be included in a poultice by itself or with charcoal, which works very effectively to remove pain.

Another herb that I have come to recommend highly for kidney and bladder infections is buchu leaf tea. Buchu was included on the first short list of medicines of the 1820 U.S. Pharmacopoeia. It is widely listed as antiseptic, diuretic, and anti-inflammatory. It is sold as shredded leaves or stems but the leaves are more potent. A tea made from the buchu leaves quickly relieves most urinary infections.

Kimberly woke up one Sunday with considerable abdominal pain and a discharge of blood in her urine. By afternoon she was very tired and sore. I gave her some buchu leaf tea to drink, and within a couple hours she was much relieved. Next morning she had a repeat episode, but again took buchu and had relief.

She also began to consciously drink more water, and seemed to be almost over her bladder infection. However, on Wednesday the problem returned, but we had run out of buchu. She found some relief with hot Sitz baths, and a hot water bottle, but by Sunday she was again in a lot of discomfort and was exhausted.

We were able to purchase shredded buchu stems, and again made some buchu tea. For several days, Kimberly drank most of her fluids as tea, and the infection was overcome.

Medicinal herbs distinguish themselves by the effects they produce on certain organs. I do not think one needs to know about all the properties of the scores of different healing herbs. It is more practical to learn how to use five or six wisely. As in the case with drugs, the concoctions of multiple herbs can quickly become confusing to the average laymen. However, used by themselves, some herbs contain properties common to other herbs, and bring relief to varied health conditions.

Take for instance spruce needles. We do not normally consider them as a healing herb but spruce needles do more than give a refreshing fragrance.

Joyce's daughter has had chronic bronchitis since childhood. With her last episode, she came home to convalesce. Joyce, a nurse, fully expected her to be homebound for several months. I suggested some dietary changes that would decrease the production of mucus. Then, as an immediate treatment to relieve the congestion that had often sparked asthma attacks in the past, I suggested boiling spruce needles. Erin was to breathe in the vapors to help decongest her lungs. We were overjoyed to hear the following week that Erin was back to work. Her doctor had tested her lungs and found them to be 95% clear—the clearest they had ever been her whole life!!

Depending on where one lives, spruce bows are free for the picking.

Having promoted a few healing herbs, a word of caution needs to be raised. Because some medicine is advertised as "herbal" or "natural" does not mean it is harmless in nature. Some herbal products are just as poisonous in their molecules as the drugs derived from them. As an informed individual you need to be aware of what you are buying. As well, otherwise benign herbs may not be as clean as we would expect them to be. Some herbal products have recently been found to be contaminated with mercury, lead and arsenic and it is not clear if it is accidental or intentional.[6] Because these heavy metals have long been associated with powerful reactions in the body, they continue to be used in some allopathic and naturopathic preparations. A safe rule of thumb is, the fewer the ingredients, the less possibility of contamination. Thankfully, healing herbs are often easily grown in our own gardens and we can be certain of their freshness and purity.

Aloe Vera

Most people are familiar with Aloe Vera as a spear-shaped tropical houseplant. Many are aware that it is also medicinal. Legend has it that Alexander the Great understood the value of its healing qualities and, on his many military campaigns, had wagons of live plants brought along specifically to treat battle wounds. Doctor Peter Atherton tells how this innocuous plant revolutionized his philosophy of medicine. "I was certainly unaware of its fabled medicinal properties and as a strictly conventional physician I had no interest in any form of complementary or alternative medicine." After twenty-eight years as a General Practitioner, with a special interest in dermatology, Doctor Atherton is currently a research Fellow at Oxford University, England, studying the medicinal effects of Aloe Vera. Referring to its many commercial preparations, he believes its simplest form is the most beneficial. "I think the product should remain as near to the natural plant as possible to achieve the correct balance of ingredients, and should be interfered with as little as possible. So, I do not favor products that have either been heat treated, filtered, concentrated or powdered." He describes it as "a tried, tested, extremely safe and non toxic remedy".[7]

Aloe Vera is specially known for its healing effects on severe burns and skin disorders, but it has also been found helpful for arthritics, asthmatics, and with the symptoms of some gastro-intestinal disorders.

Doctor Wynn shares this brief experience from the Ukraine:

> "One man came who had suffered severe second degree burns on his hand. The skin and subcutaneous tissue was sloughing off, and we could not see any good dermis tissue underneath. It really did look horrible. There were some people that had some Aloe Vera plants. They sliced the leaves lengthways to open them up. They were then cut into strips and these were laid on the man's hand with the jelly side against the hand.
>
> "The man's whole hand was covered this way, and then that was covered with a layer of gauze and then bandaged. He was sent home and returned two days later. When we removed the bandages, we saw fresh pink new skin. It looked as if it was on its way to total recovery!"

Tree Resin

Tree resin from evergreens is an ancient Native American remedy. While all evergreens have some resin, only certain trees seem to produce it abundantly. It is well known to have antiseptic and antifungal qualities, but it has also been experimented with as a healing remedy for other health problems. There are many anecdotal stories of its curative powers for cuts and infections.

Forrest worked as a woodsman for many years, but was first made aware of the benefits of tree resin by a friend who asked Forrest to get some for him. Subsequently, after years of using it in the woods for his many cuts, Forrest became confident of how it could help in healing others. But he didn't realize how much. Forrest tells of a friend who mangled her fingers in a table saw:

"The index finger was torn up down to the half moon on the nail and was not repairable when she went to the doctor. The middle finger was cut mostly off through the nail, and the doctor sewed that back on. The doctor applied the usual dressings and prescribed the usual medications. She went back in a week to have the dressings changed.

"I guess the pain of that experience was more than she wanted to face again. When it came time to go to have the stitches removed she called me instead.

"I really didn't want to get involved. She was experiencing pain, and it had been weeks since the accident. I suspected infection and preferred she go on back to her doctor. But she was adamant about my helping her. She wanted me to at least help her soak the dressing off, and see what she was looking at before she went to the doctor again.

"We spent over half an hour soaking and gently unwinding the bandages. My face must have showed my concern when I saw the mess underneath. 'It's pretty bad, isn't it?' she asked. 'Yes,' I answered. 'You need to go right up to the doctor and have this taken care of.'

"'Well, I'm not going,' she said, 'So you just take care of it for me.'

"I was looking at two fingers that were green with pus on the ends and looked like bloody, raw beefsteak clear to the wrist. The smell was just about more than one could stand. I looked at her and said, 'Friend, only if you promise to follow my instructions to the letter will I even tell you what to do. And, you must promise me that if there is no improvement in twenty-four hours you will go to the doctor.' 'Tell me what to do,' she demanded. 'I'll do it! But I won't go to the doctor again. God will help you.'

"The bone was showing out of the end of the index finger, and I told her she must make an appointment with a surgeon and have that finger reworked so she would have a good pad on the end of that bone. The flesh had all sloughed off the bone and was so infected I didn't see how she would ever have anything over that bone if a surgeon didn't make a pad. This lady was 76 years old and had poor health already. We were going to have a fight on our hands.

"We soaked her hand in hot charcoal water, alternating with cold for thirty minutes. Then I put a new dressing on the fingers. I poured tree resin from the bottle right onto the fingers and wrapped them with fresh gauze. My friend has been white with pain throughout most of this operation. I feared that at any moment she would pass out on me.

"When I finished with the dressings, she reached out for the bottle of resin and brought it nearer to read it.

"'What is this?' she queried. 'Whatever it is, I want it! How much did it cost?' She fired at me. 'I'll pay you for it right now, it took all the pain away!'

"I told her what it was, and sent the bottle home with her. We continued treatment three times a day for ten days. The raw, beaten, pulpy look all went away by the third day, and good healing set in.

"Then I left on a trip for five weeks. I was pleased with the new, pink flesh that was showing around the bone and on the middle finger, but, before I left, I warned her that she should go immediately to the surgeon and have that fingertip rebuilt. When I returned after five weeks away, she proudly showed me her hand.

"Only by looking closely could I see where the scars were. She had not gone for surgery. The finger with the bone showing had filled in completely, and most of the nail was grown back. Several years later the hand was still fine, and the finger that had been so badly mangled was not stiff. She said it tingles some at times, and is a bit numb where the nerves were damaged on the end. Otherwise, without looking closely, no one could tell that anything had ever been wrong with the fingers." [8]

Fingernail Fungus

For those of you afflicted with fingernail fungus here is a simple recipe that has worked for many. You should be able to secure the myrrh and calendula in most health food stores.

1-3 oz. myrrh
1-3 oz. calendula flowers (dry)
2 cups vinegar in a glass bottle.

Soak for three weeks, turning the bottle every day.
After three weeks, pour off the liquid, squeeze the flowers, and apply the liquid one to three times a day to the affected area, trying to keep the nail trimmed back to healthy nail, without cutting into the quick.

Pandemics

Like so many aspects of today's global society, disease is more widely circulated and shared than in previous times. The advantages of the many lines of travel and transportation have not only increased the range of goods available, but have also facilitated the spread of many diseases, and renewed the threat of pandemics. It is estimated that the Black Plague of 1347-1350 claimed twenty-five million lives, one third of Europe's population. The Spanish Influenza of 1918-1919 took twenty to forty million lives. Together these two plagues claimed almost as many lives as the two World Wars—sixty-eight million. Conservative estimates of a potential world pandemic today easily eclipse these figures.

The World Health Organization has also recognized that conventional medicine is not able to keep up with the prolific increase in certain diseases or the threat of pandemics. Realizing that new approaches must be considered, the World Health Assembly has drawn up a guideline to be promoted among member nations and not just among developing countries. There is a clarion call back to the more traditional and simpler remedies. Here are some of their acknowledgments and recommendations:

> "Aware of the accepted crucial importance of traditional medicine in many societies; recognizing the important contribution of traditional medicine to the provision of essential care; acknowledging the role of traditional medicine in the treatment of illness by informed self-medication; cognizant of the potential medical and economic value of plant substances; the WHA/WHO notes with satisfaction the progress made in the development of the program of traditional medicine; Reiterates that a substantial increase in national and international funding and support is needed to enable traditional medicine to take its rightful place in health care; urges member states to intensify activities leading to cooperation between those providing traditional medicine and modern health care, respectively, especially as regards the use of scientifically proven, safe and effective traditional remedies to reduce national drug costs; requests the Director-General:
>
> (1) to continue to recognize the great importance of this program and to mobilize increased financial and technical support as required;
>
> (2) to ensure that the contribution of scientifically proven traditional medicine is fully exploited within all the WHO programs where plant-derived and other natural products may lead to the discovery of new therapeutic substances;
>
> (3) to seek appropriate partnerships with governmental bodies and nongovernmental organizations as well as with industry in implementing this resolution;
>
> (4) to keep the Executive Board and the Health Assembly informed of the progress made in the implementation of the program of traditional medicine."[9]

These are certainly praiseworthy objectives, but ultimately, if any corporate reform, any revival of corporate responsibility for health, is ever to be successful, then it will have to begin with you and me.

In this chapter, we have related a few testimonials that demonstrate the effectiveness of simple natural remedies, but there are other simple remedies all around. The first step for all of us is to go to our local pharmacy or health-food store and purchase these items so as to have them on hand. An emergency is no time to be hurrying off to buy a fire extinguisher.

Today we have scientific studies that have endorsed the use of some traditional remedies. Today, many informed professionals are also open to exploring these time-proven remedies. We have up-to-date diagnostic tools, and many competent physicians to help identify disease. But, with Medicare programs struggling to meet increasing demands, and threats of spreading pandemics, it is time for all of us to equip ourselves to be more self-reliant. As we become more acquainted with our own physiology, and with these simple rational treatments, such as hydrotherapy and medicinal herbs, our drug bills we become smaller and smaller.[10]

Nearly a hundred years ago, John Harvey Kellogg, MD, said, "The solution to the health problems of the world today is to be found in natural remedies, not in poisoning the system with chemicals. Although they may appear to bring temporary relief, they add a debt of debilitating poison which will later result in serious problems." Nearly a hundred years later, we would do well to heed that advice. CR

---

[1] Herndon, D, editor, *Total Burn Care*, 2nd ed. New York: W B Saunders Co; 2001

[2] Lawrence, JC, British Burn Association recommended first aid for burns and scalds, Burns Including Thermal Injuries, 13:153, 1987

[3] Thrash, Agatha, MD, & Calvin, MD, *Home Remedies*, New Lifestyle Books, p. 83-84, 2001

[4] White, Ellen G, *Medical Ministry*, Review & Herald, p.230

[5] White, Ellen G, *2 Selected Message*, Review & Herald, p. 294

[6] Saper, Robert B, MD, MPH, Heavy Metal Content of Ayurvedic Herbal Medicine Products, JAMA, 292:2868-2873. Dec., 2004

[7] Atherton, Peter, MBChB, DObst, RCOG, MRCGP, Aloe Vera Myth or Medicine?, *Positive Health*, no. 20 June/July, 1997

[8] http://www.natrhealth.com/letter.html

[9] Traditional Medicine and Modern Health Care, 44th World Health Assembly, Geneva, May16, 1991, WHA44.34

[10] White, *2 Selected Messages*, p. 283

Chapter 18 Some Outlandish Name

In 1897, Ellen G. White wrote to the famed John H. Kellogg, MD, "I expect you will laugh at this; but if I could give this remedy [charcoal] some outlandish name that no one knew but myself, it would have greater influence." Then, turning her attention to the use of drugs, she went on to write, "We must become enlightened on these subjects. The intricate names given the medicines are used to cover up the matter, so that none will know what is given them as remedies unless they obtain a dictionary to find out the meaning of these names."[1]

It was Ellen White and her husband James who first recognized the potential of Kellogg when he was yet a teenager. It was the Whites who sponsored Kellogg in his medical education and worked to place him at the head of the world famous Battle Creek Hospital and medical college they helped to found. And it was the writings of Ellen White, in the area of health reform, which formed the foundation of Kellogg's great medical achievements. Who was this Ellen G. White? She is recognized by the Seventh-day Adventist Church as one of its founding members, and it is her writings in the area of health that are also credited as being the inspiration behind one of the most extensive private healthcare systems in the world.

Almost one hundred years later, in 1983, the American Library of Congress recognized Ellen White as the fourth most translated author in the world; the most translated female author in the world; and the most translated American author in the world. It is clear from perusing the volumes of material she authored, that one of her favorite topics was health reform. In 1969, the late Clive M. McCay, PhD, professor of nutrition at Cornell University, wrote: "In spite of the fact that the works of Mrs. White were written long before the advent of modern scientific nutrition, no better overall guide is available today."[2] Apart from nutrition, what else did this prolific visionary have to say in the area of health reform that relates to drugs versus simple home remedies?

The following statements represent the principle Ellen White advocated. She believed that simple natural remedies should be our first choice in treating disease, as we learn to discard those things that are injurious to our health.

Drugs

At a time not all that different in some ways than ours, when the sick and ailing were constantly being bombarded with information about some new patent wonder drug, Ellen White advocated caution and common sense:

> "Thousands who are afflicted might recover their health if, instead of depending upon the drugstore for their life, they would discard all drugs, and live simply, without using tea, coffee, liquor, or spices, which irritate the stomach and leave it weak, unable to digest even simple food without stimulation."[3] (1903).

Considering this statement a hundred years later, it truly is visionary. When Ellen White penned these words at the beginning of the 1900s, much of the medical profession was divided between the business of drugging, quackery, and advocating every conceivable remedy. Today, a hundred years later, judging from the number of different drugs being introduced and the different therapies being advertised, things are certainly no less confusing. Because packaging and advertising have become so much more sophisticated, the potential for being seduced with some inappropriate or downright dangerous drug or therapy has been magnified a hundred times.

Considering the plethora of treatment programs around today, Ellen White's words of caution a hundred years ago are even more relevant in this new century than they were in the last one. Speaking to the members within her church, to both physicians and laymen, she sought to remind them of a more rational way of healing than the doling out of drugs:

> "There are many ways of practicing the healing art, but there is only one way that Heaven approves. God's remedies are the simple agencies of nature that will not tax or debilitate the system through their powerful properties. Pure air and water, cleanliness, a proper diet, purity of life, and a firm trust in God are remedies for the want of which thousands are dying; yet these remedies are going out of date because their skillful use requires work that the people do not appreciate. Fresh air, exercise, pure water, and clean, sweet premises are within the reach of all with but little expense; but drugs are expensive, both in the outlay of means and in the effect produced upon the system."[4] (1885).

Recognizing that there were gross sanitation problems in her day and acknowledging the increase of pollution in ours, the fact remains that if people will make their health a priority, then the vital elements she names are still within reach. If the money spent on useless trifles, lotteries, health-destroying habits, and quick-fix drugs, was instead carefully saved, there would be sufficient means for even the poor to access the heaven-sent remedies of sunshine, clean air and water.

In Ellen White's day, charcoal did see some experimental use as a natural remedy for several conditions, but it is evident the world had yet to catch on. Sad to say, apart from the industrial and environmental sectors, charcoal is still a relatively well-kept secret. One company that began to explore the potential of charcoal near the end of Ellen White's career is now one of the world leaders in charcoal research and development. It produces over 150 grades of charcoal for every imaginable application. Unfortunately, in the field of health, as we have seen, the extensive application of charcoal is largely behind the scenes.

I freely admit that I would know little or nothing today about the efficacy of charcoal if it were not for the writings of Ellen White. It was her writings that had attracted the attention of Dr. Graves in Guatemala, and it was his very practical enthusiasm that introduced them to me. If those I studied under had not themselves read Ellen White's recommendations and experimented with charcoal, I seriously doubt I would have ever gained a working knowledge for myself. Of the anecdotal experiences with charcoal that I collected, for most, the inspiration to try charcoal may be traced back to Ellen White's published stories. While some of her contemporaries were using charcoal, Ellen White is the only health pioneer to have left a written log of her varied experiences. It is for that reason that I have included this chapter.

This project on charcoal is far too narrow to capture how truly farsighted and innovative Ellen White's experiences and writings were. For those interested in exploring more of her insights into complete health, two of her works are cited in the list of **Further Reading** in the Appendix. The rest of this chapter is a collection of her experiences involving charcoal, showing her absolute faith in its potential.

Charcoal

In her late 60s, Ellen White traveled across the Pacific Ocean to help found a college in the wild Australian countryside. Writing to Dr J. H. Kellogg, she extols the virtues of charcoal:

> "One of the most beneficial remedies is pulverized charcoal, placed in a bag and used in fomentations. This is a most successful remedy. If wet in smartweed boiled, it is still better. I have ordered this in cases where the sick were suffering great pain, and when it has been confided to me by the physician that he thought it was the last before the close of life. Then I suggested the charcoal, and the patient slept, the turning point came, and recovery was the result. To students when injured with bruised hands

> and suffering with inflammation, I have prescribed this simple remedy, with perfect success. The poison of inflammation was overcome, the pain removed, and healing went on rapidly. The most severe inflammation of the eyes will be relieved by a poultice of charcoal, put in a bag, and dipped in hot or cold water, as will best suit the case. This works like a charm."[5]

One is left to wonder, if the attending doctor had sought advice earlier, how much more suffering may have been avoided?

Typhoid Fever

As avid a health reformer as she was, Ellen White recognized the need for wisdom and prudence in dealing with the sick. Notice this experience and counsel to a headstrong reformer:

> "Dear Brother Peter: I still remember another case. At our first camp-meeting here, held in Brighton, a young lady was taken sick on the ground, and remained sick during most of the meeting. She was thought to have typhoid fever, and although many prayers were offered in her behalf, she left the ground sick. Dr. M. C. Kellogg, half-brother to J. H. Kellogg of Battle Creek, was attending her. He came to me one morning, and said, Sister Price is in great pain. I cannot relieve her. She cannot sleep, and every breath seems as though it would be her last. We prayed for her, and then like a flash of lightning there came to me the thought of the charcoal. "Send to the blacksmith for charcoal, and pulverize it," I said, "and put a poultice of it on her side." He tried this, and in one hour he came to me and said, "That prescription was an inspiration from God. Sister Price could not have lived until now if no change had come." The sick one fell into a restful sleep; the crisis passed, and she began to amend… and is alive and well today.
>
> "All these things teach us that we are to be very careful lest we receive radical ideas and impressions. Your ideas regarding drug medication, I must respect; but even in this you must not always let the patients know that you discard drugs entirely until they become intelligent on the subject. You often place yourself in positions where you hurt your influence and do no one any good, by expressing all your convictions. Thus you cut yourself away from the people. You should modify your strong prejudice."[6] (1899).

In admonishing "Brother Peter", Ellen White reminds those who are zealously reform minded, that little is gained from advocating progressive health principles until a proper foundation is laid. Truth, like the dawning of the day, is progressive in the sense that it is an active working principle, and it is naive to expect that the uninformed will automatically embrace practices that are new and strange. Today we are a culture of pill poppers. There is a ground swell of curiosity and interest in simple remedies, but that does not mean people will automatically try them without some introduction to the more basic principles they are founded upon. This requires more teaching and not just treating on the part of health practitioners. As people become more and more knowledgeable about their own anatomy and physiology, they will be more inclined to make use of the simple remedies. And, as Ellen White noted, "their drug bill will become less and less."

Inflammation and Pain

In the following letter to the director of the then new Loma Linda Hospital in southern California (one of several hospitals she helped to develop), Mrs. White relates her initial introduction to charcoal:

> "I have ordered the same treatment for others who were suffering great pain, and it has brought relief and been the means of saving life. My mother had told me that snakebites and the sting of reptiles and poisonous insects could often be rendered harmless by the use of charcoal poultices. When working on the land at Avondale, Australia, the workmen would often bruise their hands and limbs, and this in many cases resulted in such severe inflammation that the worker would have to leave his work for some time. One came to me one day in this condition, with his hand tied in a sling. He was much troubled over the circumstance; for his help was needed in clearing the land. I said to him, "Go to the place where you have been burning the timber, and get me some charcoal from the eucalyptus tree, pulverize it, and I will dress your hand." This was done, and the next morning he reported that the pain was gone. Soon he was ready to return to his work."[7] (1908)

Ellen White goes on to remind her reader that God has not abandoned us without "simple remedies which when used will not leave the system in the weakened condition in which the use of drugs so often leave it." She emphasizes the need for "well-trained nurses who can understand how to use the simple remedies that nature provides for restoration to health, and who can teach those who are ignorant of the laws of health how to use these simple but effective cures."

This clear, succinct practical counsel of treating and teaching, given a hundred years ago, is still waiting to be implemented in the majority of medical schools of this new century.

Bloody Dysentery

Here, Ellen White describes a much earlier experience as she, her husband and others, prepared to drive a herd of horses north from Texas to Colorado:

> "A brother was taken sick with inflammation of the bowels and bloody dysentery. The man was not a careful health reformer, but indulged his appetite. We were just preparing to leave Texas, where we had been laboring for several months, and we had carriages prepared to take away this brother and his family, and several others who were suffering from malarial fever. My husband and I thought we would stand this expense rather than have the heads of several families die and leave their wives and children unprovided for.
>
> "Two or three were taken in a large spring wagon on spring mattresses. But this man who was suffering from inflammation of the bowels, sent for me to come to him. My husband and I decided that it would not do to move him. Fears were entertained that mortification had set in. Then the thought came to me like a communication from the Lord to take pulverized charcoal, put water upon it, and give this water to the sick man to drink, putting bandages of the charcoal over the bowels and stomach. We were about one mile from the city of Denison, but the sick man's son went to a blacksmith's shop, secured the charcoal, and pulverized it, and then used it according to the directions given. The result was that in half an hour there was a change for the better. We had to go on our journey and leave the family behind, but what was our surprise the following day to see their wagon overtake us. The sick man was lying in a bed in the wagon. The blessing of God had worked with the simple means used."[8]

As we have seen in previous chapters charcoal continues to be a powerful aid in the treatment of all forms of diarrhea.

Insect Bite

Ellen White recorded many of her more memorable experiences with charcoal while pioneering in Australia. She recounts in her diary, how, in spite of their lack of medical facilities, they were still able to help one suffering child, using a simple charcoal poultice.

Of that experience she wrote:

> "The building work on our hospital has not yet commenced, but the land is being cleared preparatory to building. We need a hospital so much. On Thursday Sister Sara McEnterfer* was called to see if she could do anything

for Brother B's little son, who is eighteen months old. For several days he has had a painful swelling on the knee, supposed to be from the bite of some poisonous insect. Pulverized charcoal, mixed with flaxseed, was placed upon the swelling, and this poultice gave relief at once. The child had screamed with pain all night, but when this was applied, he slept. Today she has been to see the little one twice. She opened the swelling in two places, and a large amount of yellow matter and blood was discharged freely. The child was relieved of its great suffering. We thank the Lord that we may become intelligent in using the simple things within our reach to alleviate pain, and successfully remove its cause."[9] (1899) (*A trained nurse of experience, well qualified for this type of service, who accompanied Mrs. White, and assisted her both as a traveling companion and private secretary.)

Inflamed Knee

Ellen White refers to another case that was brought to their primitive school in Australia. Once again conventional medicine had failed to help, and the case had been diagnosed as incurable. The reader will notice charcoal was used successfully in conjunction with hydrotherapy.

Here is an excerpt of an article sent to a church magazine in America:

"One of the boys who came with the father was a cripple, using crutches, and he cooked while the others worked. This boy is thirteen years old, and had been troubled with a knee-swelling for five years. For eleven months he was confined to his bed under the care of a physician. Sister McEnterfer had treated him with water compresses and pulverized charcoal, until the inflammation had been relieved. He was so much better that he laid aside his crutches, and attended to the cooking, as has been mentioned. But this was too much, and the knee troubled him again. It was necessary to give him a thorough course of treatment, so we took him into my own house and gave him constant care. There was a large swelling under the knee, which he called his 'egg'. This swelling was opened and discharged freely, and from it were taken pieces of bone.

"What power there is in water! He improved rapidly, and he was given light work—copying letters in the letter-book, learning to write on the typewriter and other things. We now send him to school. We board and clothe him and his father pays his tuition. We keep him for the benefit we may do the boy and he is good material to work upon. The father and mother cannot express their gratitude; for physicians, who had previously examined and treated the boy, had told them that he would be a cripple for

> life. The parents now look upon the boy—active and healthy—and you can judge how they feel. This is our field for missionary work …
>
> "This boy is the third case of terribly injured limbs which have been cured by simple remedies. In each case they have been pronounced incurable by physicians. These cases have been maltreated, and it was thought that blood-poisoning had set in, in two cases. Sister McEnterfer took these cases and treated them with great pains-taking effort for weeks."[10] (1899).

This next brief word of instruction was clipped from another letter written while in Australia. It was addressed to friends not very far away, who were trying to help a family member suffering from a debilitating fever. Ellen White is sending the charcoal with her personal nurse who will stay to help the family:

> "I send you at this time pulverized charcoal. Let him drink the water [mixed with charcoal] after it has stood a while to extract the virtue. This should be cold when used. When used for fomentations over the bowels, the coal should be put into a bag, sewed up, and dipped in hot water. It will serve several times. Have two bags; use one and then the other."[11] (1897).

While charcoal will work without hydrotherapy, used in combination with warm moist applications, the effect is very soothing upon a distraught mind.

Eye

Recognizing the delicate nature of the eye and the dangers of some common treatments, Ellen White writes to a friend of her own experience of severely injuring her one eye:

> "My left eye gives me considerable trouble. It has been painfully weak, and the pain in my cheekbone has caused me much anxiety. I think I told you that about a year ago I had a fall. I was filling a tin milk-pan with oranges for brother McCann, and when I rose from my stooping position, and attempted to walk, I saw that I was falling face downward, into a pile of small, sharp stumps, which had been brought there for fuel. I darted a prayer to heaven, and fell heavily forward. The milk pan struck the stumps with great force, and when I fell, it struck me just beneath the eye. The pan was bent almost double. Brother McCann came and lifted me up.
>
> "After, I had a very painful time with my whole head. My cheek was swelled large and hard. Ella White was with me at the time, and she pounded up charcoal for me. With this pulverized charcoal and hot water I treated my face for hours, till the soreness and pain was killed."[12] (1896)

Whereas the use of charcoal brought relief from the intense pain, as she wrote, the injury was still causing her "considerable trouble" and anxiety a year later. It is clear she felt her initial treatment had served her well, or she would not have related it as she did. Considering that she was sixty-nine years old, the severity of the fall (bending the pan nearly in two), the trauma to her face, and the primitive circumstances, it is surprising that she would continue to recover. Her dedication to the school is also amazing. Most people would have quickly packed up and bought a one-way-ticket home. However, she was rewarded for her dedication.

Only those who have traveled abroad, and lived and worked in such primitive circumstances, can appreciate the temptation there is to resort to some quick-fix pain-reliever. In order to relieve some present anxiety or pain, many people resort to some narcotic, only to set themselves on a path of lifelong addiction. Ellen White understood the poisonous and/or addictive nature of drugs. So instead of using them, she continued to apply the simple remedies, confidently knowing they would do no harm, as they slowly but surely helped the natural vitality to recover. She would later write that her health was "better than it had been in years".

Two years after her fall, looking back on her experience, she was even more convinced of her decision. She wrote emphatically:

> "Medical practitioners have experimented on the eyes, and in so doing, have not only forever weakened the strength of the delicate organs, but the injury done has extended to the brain, through the nerves connecting the eye with the brain. I have positively refused to use anything but hot water with a little salt, or pulverized charcoal put in a bag, and dipped in hot or cold water, as is most agreeable. Let there be no meddling with the eye. Use only the most mild applications."[13] (1898)

Headaches

The reader will see in this next story how simple remedies are employed in conjunction with a restricted diet. Here is a case where, "with but little expense", an invalid finds rest and healing away from the city traffic. Again, Ellen White brings the sick into her country home for treatment. She writes to a friend:

> "Sister Harlowe was paying twelve shillings a week for room and board in the city. I brought her to my country home, and she has lost her headache. We have had her now about two months. I ask her nothing for board. I want her to get well.
>
> "Sister Harlowe has been an invalid for sometime, in consequence of her internal organs fastening to the backbone. She has been proprietor of a large dressmaking establishment. In this business her difficulties developed.

> Her head suffered because of stomach difficulties and the trouble I have mentioned...
>
> "Here she rides on horseback and does a little sewing. She is improving in health so much that we have hopes that she will be able to educate a class in school in the science of sewing properly. This is the one who was advised to eat meat and butter. She uses neither. She has plenty of milk and cream from cows which we keep in as healthy a condition as possible. I am raising my own stock. Sister Harlowe has not had any kind of medicine but charcoal soaked in water. She drinks this water, and is doing excellently without meat, butter, tea or coffee."[14] (1898)

Here we see in a practical way, what Ellen White termed as "God's remedies", being applied as a package, fresh air and sunshine, simple exercise, rest, a simple diet along with a simple home remedy.

Trust

When we are sick, we want to be able to have confidence in man, but experience has taught us that there is really only One we can trust in completely. As well, Ellen White points out that we need to know ourselves. We do well to recognize that each person is unique and that their treatment, including diet, may need to be tailored to their unique condition. Writing to a chronically ill person, Ellen White stresses the importance of mental health as it relates to total health:

> "I would advise that the charcoal compress be worn only occasionally. If you should drink charcoal water, it would not do you any harm. The charcoal itself may be a little irritating to a stomach as sensitive as I judge yours is. Use your own judgment, and trust in living faith to God as you work. You must come to the position where you realize that the Lord does not want you to be sick. The Lord would have you well. Make up your mind to be well. Do not eat vegetables but make arrangements to have the best homemade bread and fruit. Do not taste of any bread that is sweetened. You must be careful, and then trust in God, believing that He wants you to be healed. He is the only true, unerring physician in the world. He loves us, and wants us to be well and happy."[15] (1898)

Someone once calculated for me that twenty pounds of "do" equals one pound of "done", which is the motivating force for twenty pounds of "do". Physical health promotes mental health. They in turn promote spiritual strength which is the real powerhouse behind physical and mental health.

Indigestion

Writing to those having to do with directing a hospital and training nurses, Ellen White emphasizes the importance of cultivating the mental powers, both for the patients and for the caregivers. She always directs the focus away from costly equipment to those remedies that are accessible to all. Here she relates once again some of her experiences in Australia:

> "Fresh air and sunshine, cheerfulness within and without the institution, pleasant words and kindly acts—these are the remedies that the sick need, and God will crown with success your efforts to provide these remedies for the sick ones who come to the sanitarium. By happiness and cheerfulness and expressions of sympathy and hopefulness for others, your own soul will be filled with light and peace. And never forget that the sunshine of God's blessing is worth everything to us.
>
> "Teach nurses and patients the value of those health-restoring agencies that are freely provided by God, and the usefulness of simple things that are easily obtained.
>
> "I will tell you a little about my own experience with charcoal as a remedy. For some forms of indigestion it is more efficacious than drugs. A little olive oil into which some of this powder has been stirred tends to cleanse and heal. I find it is excellent. Pulverized charcoal from eucalyptus wood, we have used freely in cases of inflammation …
>
> "When we first went to Cooranbong [Australia], the men who were clearing in the woods often came in with bruised hands. In these and other cases of inflammation, I advised the trial of a compress of pulverized charcoal. Sometimes the inflammation, which was very high before the compress was applied, would be gone by the next day.
>
> "Always study and teach the use of the simplest remedies, and the special blessing of the Lord may be expected to follow the use of these means which are within the reach of the common people."[16] (1903)

The reference to using eucalyptus wood for charcoal is likely due to its availability and hardness. The denser the wood, the better a quality of charcoal you will get. Elsewhere she makes mention of eucalyptus oil mixed with honey and taken orally as a cough medicine. She writes concerning a woman suffering a chronic cough:

> "I cannot advise any remedy for her cough better than eucalyptus and honey. Into a tumbler of honey put a few drops of the eucalyptus, stir it

up well, and take whenever the cough comes on… If you will use this prescription, you may be your own physician. If the first trial does not effect a cure, try it again. The best time to take it is before retiring."[17] (To a worker).

Bad Breath

As beneficial as exercise and purposeful labor are for us, they can become our enemy. This next letter from Ellen White to Brother J seems self-explanatory:

"But Brother J's health has been sacrificed to earnest work, continuous labor. In his zeal and earnestness he has lost sight of self. He wishes to converse with me, but what an offensive breath he had. I tried not to inhale it, but I think I did, for I was taken sick like one poisoned. I talked with him seriously, and told him that he had been doing great injury to himself…

"I think much of Brother J, but he has made a mistake in feeling that he must do all the work he could possibly accomplish. At times he would become so weary that he could not eat as he should, and would take food that was wholly unfitted for his wearied condition. He kept late hours, and often did not get rest before eleven or twelve o'clock at night. This irregularity was seriously felt by his wife. She became nervous, and was losing her health and vitality. I showed him that these habits had been all wrong, that he must call a halt. I told him it was necessary for us to know ourselves. I said, you are now a sick man, and you feel your need of a physician. We are not wise in the knowledge of others, but we know that in our own individual selves there are great deficiencies.

"The wisest have to learn their lessons by patient experience. You have been unaware how weak and unwise you are. Now you must be made to feel that you have not treated yourself as you should, but have disregarded the laws of health by your terrible neglect of yourself. Something must be done. That offensive breath must be purified. Get pulverized charcoal, soak it in water, and drink this water freely. Eat no vegetables. Eat fruit, and plain well-baked bread. Take light exercise, and at night, wear a charcoal compress over the liver and abdomen."[18] (1898)

As with the one earlier, this reference to leaving vegetables out of the diet, does not contradict her praise of vegetables as part of a healthy menu, but rather shows her insight into the need for therapeutic diets. Given that vegetables are generally harder to digest, and considering the man's symptoms suggest indigestion, this recommendation seems very practical.

Poisons

In this next piece from a letter to a co-worker, Ellen White gives us a window into how comprehensive the work of a health giver can, and should be:

> "Sister McEnterfer is nurse and physician for all the region round about. She has been called upon to treat the most difficult cases, and with complete success. We have at times made our house a hospital, where we have taken in the sick and cared for them. I have not time to relate the wonderful cures wrought, not the dosing with drugs, but by the application of water. We use charcoal largely, making it into poultices. It destroys the inflammation, and removes the poison. We are teaching the ignorant how to become intelligent and keep well."[19] (1899)

Here we see that the art of healing does not end with nursing the sick back to health, but also includes teaching proper health principles.

Mixed with Sweet Oil

In this last reference to charcoal, Ellen White is also seeking to emphasize personal responsibility for becoming informed on all that pertains to our health:

> "Get from the druggist some pulverized charcoal, and use it freely. Mix it with sweet oil. Thus it can be taken with less difficulty than mixed with water. I think that you would obtain benefit from the use of charcoal compresses, pulverized charcoal moistened, put in a flannel bag, and placed over the affected part. When my husband was sick, I had recourse to many remedies, and I know the value of charcoal as a healing agency. I have worked for my husband with marked success when his life was in grave peril. I did not want a physician for him; for I knew that I had tact and skill, and that with faith in God I could be his physician."[20] (1904)

Today you will find that the informed conscientious pharmacist will also have charcoal available for his customers. Ellen White took time, not only to inform herself about her body, but also to experiment with different simple remedies that were available wherever she happened to be. Because of this willingness to learn, she believed herself to be fully capable of treating most cases of illness. She did on occasion seek out a physician's help, but her belief was that generally, our need for a doctor should be no more frequent than our need for a lawyer. If we follow the principles of health she advocated, and employ nature's simplest remedies with the same care and trust in God as she did, then we too should expect similar results, including fewer and fewer visits to a doctor.

Jethro Kloss

Jethro Kloss was one individual who, after stumbling upon the writings of Ellen White, saw his visits to a doctor come to an end, and went on to become a world famous health reformer in his own right.

Jethro Kloss writes in his magnum opus, *Back To Eden*:

> "God has provided a remedy for every disease that might afflict us. Satan cannot afflict anyone with any disease for which God has provided a remedy. Our Creator foresaw the wretched condition of mankind in these days, and made provision in nature for all the ills of man. If our scientists and medical colleges would put forth the same effort in finding the virtues in the 'true remedies' as found in nature for the use of the human race, then poisonous drugs and chemicals would be eliminated and sickness would be rare indeed. If they would make use of only these remedies that God has given for the 'service of man' it would bring an untold blessing to the world. In these distressing days, the use of a simple natural diet would prevent much suffering and save money. The most important subject for people to study, should be: How can we live our allotted time without suffering? God has surely made this possible."[21] (1939)

Jethro Kloss' classic book, which has sold over five million copies, was written as a result of having become acquainted with Ellen G. White's books on health. After leaving home at an early age, he lived in boarding houses and hotels, eating little except devitalized foods, until his health began to fail.

He suffered a complete breakdown, and became so weak, that at one time he was in bed for three months with pain so severe he wrote, "No tongue can describe. Death would have been a welcomed release."

He tried every remedy known to the medical profession, but none of his doctors gave him any hope. He had no thought of ever getting well again, and had made his will. It was then that he came across the books written by James and Ellen White. There he read, "Nine out of ten would get well if they used simple God-given means.... Do not eat food robbed of its life-giving elements."

"This was an entirely new line of thought to me", he said. He recalled that when he was young, his parents were called to a neighbor's house after someone had been given up to die. His parents, with their herbs, fruit juices, and vegetable broths, helped the sick person back to health. He began to accept many of the suggestions in Mrs. White's books until he was once again a well man.

It was not long after that he had opportunity to help a man suffering from typhoid fever. The man's doctor had said he would die:

"I began to treat the man with the simple remedies that I had found in the books of Mrs. E. G. White. I stopped all poisonous drugs and gave him plenty of water and juices, and used other simple means, such as hot and cold fomentations, which proved a great means of allaying his pains. In about six weeks the man was up and around."

Rational Methods

Once again, writing to those directing the medical institutions of the church, to doctors, and to nurses, Ellen White champions the simple remedies of nature over the debilitating effects of drugs. Recalling her own experiences, she calls upon those pledged to "do no harm", to exchange their poisonous drugs for the more benign remedies that God has promised to bless:

"Those God-fearing workers will have no use for poisonous drugs. They will use the natural agencies that God has given for the restoration of the sick. Time and again I have told the workers in our sanitariums [hospitals] that from the light that God has given me I know that they need not lose one patient suffering from a fever, if they take the case in hand in time and use rational methods of treatment instead of drugs.

"My husband and I were neither doctors nor the children of doctors, but we had success in the treatment of disease. In a time when many of the people—even the children of physicians—were dying all around us, we went from house to house to treat the sick, using water and giving them healthful food. Through the blessing of God, we did not lose a single case... Oh, how great are the possibilities that He has placed without [archaic for 'at'] our reach! He says:

"'Whatsoever ye shall ask the Father in My name, He will give it you.' He promises to come to us as a Comforter to bless us. Why do we not believe these promises? That which we lack in faith we make up by the use of drugs. Let us give up the drugs, believing that Jesus does not desire us to be sick, and that if we live according to the principles of health reform, He will keep us well."[22]

Arguably, Ellen White is one of the most notable and credible health reform writers of the late 19th and early 20th centuries. Her prolific writings in the area of health are credited as laying an integrated foundation of health principles that have resulted in over two hundred international studies. These private and government sponsored research projects have noted the significantly higher longevity scores of seven to eight years, compared to the general American population, amongst those within her faith persuasion who practice the principles she advocated.

In this 21st century, we can only lament what great advances healthcare might have realized if we had more fully explored the rational treatment of disease without drugs. Dr. J. H. Kellogg took hold of the progressive writings of Ellen White on dietary reform, and brought world fame to the Battle Creek Hospital in Battle Creek, Michigan. At the same time, W. K. Kellogg, as an outgrowth of his work with his brother at the Battle Creek Hospital, founded the Kellogg's® international food company. The College in Australia, that Ellen White helped to found, also became the home of Sanitarium Foods®, the largest breakfast food distributor in that quadrant of the world. Jethro Kloss, after years of study and experimentation, became one of the pioneer developers of the still expanding soybean food industry. He also published *Back to Eden* and its companion cookbook, which, seventy years and six million copies later, continue to be bestsellers.

Imagine, if America had produced another such visionary to take hold of charcoal's potential, how many more new and exciting charcoal products would the world be enjoying, especially in the area of health? CR

---

[1] White, Ellen G, *2 Selected Messages*, Review & Herald Pub. p. 294, 1958
[2] McCay, Clive M, *Review and Herald*, February 26, 1969
[3] White, *2 Selected Messages*, page 291
[4] White, Ellen G, *Testimonies Vol. 5*, Review & Herald Pub. p. 443
[5] White, *2 Selected Messages, page 294*
[6] The Paulson Collection of Ellen G. White Letters, page 27
[7] White, *2 Selected Messages*, pp 295-296
[8] *Ibid.*, p. 299 Letter 182, 1899
[9] White, Ellen G, Manuscript 68, 1899. EG White Diary, April 25, 1899
[10] White, Ellen G, *The Gospel Herald*, October 1, 1899
[11] White, Ellen G, Manuscript Releases 20, page 279, 1897
[12] White, Ellen G, Letter 119, 1896 - Not indexed
[13] *Ibid.*, Letter 37, 1898 - Not indexed
[14] *Ibid.*, Letter 84, 1898 - Not indexed
[15] *Ibid.*, Letter 92, 1898 - Not indexed
[16] The Paulson Collection of Ellen G. White Letters, Letter 100, 1903, page 37-38
[17] White, Ellen G, Letter 348, 1908
[18] White, Ellen G, Letter 115, 1898 - Not indexed
[19] *Ibid.*, Letter 74, 1899 - Not indexed
[20] *Ibid.*, Letter 75, 1904 - Not indexed
[21] Kloss, Jethro, *Back to Eden*, Lotus Press, Wisconsin, p. III, 1989
[22] White, Ellen G, Manuscript Release 19, p.51

Chapter 19 A Family Practice

"If one has to be wealthy to be healthy, then God only loves rich people." "If the means to achieve and maintain optimum health is within the reach of only a select few, then God is elitist." I have wrestled with these thoughts over the years and over the miles. Are they true?

Since my first experience in Guatemala with charcoal and simple natural remedies, I have traveled and worked in several foreign countries, and I have often been struck by the tremendous disparities between America and those developing countries. The availability and degree of medical care is one of the most obvious differences. Western countries are wealthy and developing countries are poor, but if God is no respecter of persons, if "God is love", and God loves all, then health and healing must be as accessible to the poor and uneducated as it is to the wealthy and the educated. God's outstretched hand of healing should be no more than an arm's length away.

Nepal

There I was working with a small group of volunteers at a remote clinic in Nepal, beyond the reach of electricity and telephone lines. A four to five hour hike, from the nearest main road to civilization, up verdant valleys and over hills that would humble mountains in other lands, would bring you to this small medical oasis. It did not take any of us long to realize just how vital the small clinic was to the numerous surrounding villages, and those situated even farther from the main road. From the main road it could be a torturous bus ride to the closest hospital a day away. Clearly the people needed medical help closer at hand. We decided we would employ eight doctors, and make them available at a price the poorest could afford.

Before I introduce this family of doctors "without borders", let me first diverge a bit. Over a hundred years ago, pioneer health reformer and lecturer Ellen White advocated, "Always study and teach the use of the simplest remedies, and the special blessing of the Lord may be expected to follow the use of these means which are within the reach of the common people."[1]

In her classic book on health, *The Ministry of Healing*, Ellen White elaborates:

"Pure air, sunlight, abstemiousness, rest, exercise, proper diet, the use of water, trust in divine power—these are the true remedies. Every person should have a knowledge of nature's remedial agencies and how to apply them. It is essential both to understand the principles involved in the treatment of the sick and to have a practical training that will enable one rightly to use this knowledge.

"The use of natural remedies requires an amount of care and effort that many are not willing to give. Nature's process of healing and upbuilding is gradual, and to the impatient it seems slow. The surrender of hurtful indulgences requires sacrifice. But in the end it will be found that nature, untrammeled, does her work wisely and well. Those who persevere in obedience to her laws will reap the reward in health of body and health of mind."[2]

The use of simple remedies, accessible and affordable, these were the medical qualifications we deemed vital in our team of doctors.

Eight Doctors

Where could we find a medical team that we could make accessible and affordable to the poor peasant farmers that populated the remote district of Parbat? We finally decided on the eight-member family practice we were all familiar with. Any health care system can afford them without stressing exhausted budgets. In fact this family of doctors travels and practices internationally and often go by the acronym NEWSTART®.

So who are these doctors? May I introduce to you Dr. Nutrition, Dr. Exercise, Dr. Water, Dr. Sunshine, Dr. Temperance, Dr. Air, Dr. Rest, and Dr. Trust in Divine Power, ready to take your call any time of the day or night. Here we have the eight doctors that we employed to help restore the sick and suffering we met every day.

While they promote health and healing, they also emphasize prevention. This strategy for introducing preventive medicine to the local people proved to be very practical and popular. For people who are marginalized, especially from health care, having skills they can use to help themselves gives them a sense of empowerment. It gives them tools and strategies they can use right where they are, without the heavy expense of lost work time, travel, and costly drugs.

Restored health is the goal of so many people today, either for themselves or for a loved one, for a patient, and, for many people, their bank account. But, treating disease has become big business from research scientists right through to pathologists and morticians, from pharmaceutical giants right through to stockbrokers and drugstores.

However, studies have shown that adherence to just seven health practices would add six to eight years to one's life, while saving developed nations eighty percent of their already crippling health budgets.[3]

What are these seven health practices?

- Sleep, seven to eight hours daily
- Eat breakfast almost every day
- Never or rarely eat between meals
- Being at or near one's height adjusted weight
- Regular physical activity
- Not smoking cigarettes
- Moderate or no use of alcohol

As you can see, these seven health practices fall in line with the NEWSTART® team strategy of disease prevention. For all the enchantment with drugs, MRI's and proton accelerators, the reality is that most disease can be intelligently treated right within the home.

I chose to include this chapter on preventive health because I know by experience that any remedy, including charcoal, will not work indefinitely if there is no corresponding change in one's lifestyle. I do not promote charcoal as a daily supplement for the rest of one's life, even though it may greatly benefit some as a general detoxifier for a couple of months. As far as managing disease is concerned, charcoal is, at best, only a steppingstone. We need to progress to the next step of prevention.

In today's frantic pace, our tendency is to shift our own responsibility for maintaining and restoring health onto the shoulders of others. This goes for our spiritual, mental, emotional, as well as physical health. We should be the gatekeepers of our total health. When we give the control of our health to others, our health comes at an ever-increasing price. The more we depend on others, the more helpless we become, the more we have to pay out in cash, and consequently the more we pay out in anxiety with all its spin offs.

Is there a state or nation today that does not struggle to meet the burgeoning financial drain of health care on its economy? Politicians are driven to make pledges that they can in no way honor. At the same time many health professionals are burning out, as evidenced by their high drug addiction and suicide rates.[4] High wages with a host of perks no longer appeal to many workers employed in the medical industry. Many professionals realize there is no compensation that can match the emotional, mental, and physical pressures that drain them. A *New*

England Journal of Medicine article (1998) noted that among American physicians, "Overt and vocal dissatisfaction with their lot is quite common and almost routine… There has been an undercurrent of unhappiness among physicians for many years, but the complaints seem more widespread and more strident now."[5] Clearly a new philosophy is needed. Today we have abundant light pointing to a better way.

Dr. Nutrition

Thomas Edison, the inventor of the light bulb, had a vision for tomorrow. He wrote, "The doctor of the future will no longer treat the human frame with drugs, but rather will cure and prevent disease with nutrition." Perhaps he had read Hippocrates, who said 2,400 years earlier, "Let food be thy medicine, and let thy medicine be food". We need to take responsibility for what we eat, how much and how often. Since the beginning of the 20th century, there has been a steady decline in the consumption of whole grain foods and vegetables, which are high in fiber. At the same time there has been a dramatic increase in the consumption of animal products and processed foods that are low in fiber. It is no wonder constipation, with all its spin-off ailments, has become so prevalent.

Along with an increase in meat and refined food consumption, there has been a rapid rise in obesity, heart disease, and cancer. One doesn't have to be a nuclear physicist to make the connection. But, back in the 1950s, Albert Einstein did recognize this trend. His solution? Even though his field of expertise was not nutrition, he wrote, "Nothing will benefit human health and increase the chances for survival of life on earth as much as the evolution to a vegetarian diet." Of course being a vegetarian doesn't mean one is free from the temptation to indulge carelessly. Unfortunately, whether carnivores or vegetarians, Americans have a sweet tooth.

To "celebrate" their passage into the new millennium, the average American unwittingly consumed over 150 pounds of sugar in 1999. This statistic, coupled together with fast food and snack consumption across the U.S., and the rapid rise of diabetes, prompted pharmaceutical giant Eli Lilly, in 2003, to build the largest factory of its kind in America to produce just one product—insulin. Massive insulin production, or the saliva of Gila monsters, is not the solution for diabetes. Eli Lilly, and other pharmaceuticals, are not charitable organizations, they are big business. Better food choices are needed, not more drug choices. So beware, "When you see the Golden Arches you are probably on the road to the Pearly Gates."[6] William Castelli, MD—director of the world famous Framingham Heart Study.

Some chronic disease sufferers, after grazing on their cravings, try to solace their conscience by emphasizing the role of genetic inheritance. They say, "It's in my genes." It is absolutely true, inheritance does affect our health, but while heredity loads the gun it is lifestyle that pulls the trigger.

Instead of greasy fried foods, fibreless refined foods, and empty calorie foods, Dr. Nutrition's prescription for a balanced menu emphasizes those foods first given to man in the Garden of Eden:

- Whole grains—both in breads and well-cooked cereals.
- Fresh fruit in abundance.
- Nuts in all their shapes.
- Fresh vegetables when in season, or preserved.
- Variety as the spice of choice, rather than irritating condiments.

This menu planner only includes three meals a day: breakfast like a king; lunch like a prince; and supper like a pauper. Nothing more until you break your fast in the morning, and nothing between meals except water or herb teas.

Dr. Exercise

Four large population studies in the U.S. and Canada have shown that the feeling of general well being is somewhat greater, and depression is much less frequent, in those who take frequent exercise relative to those who take little or no exercise.[7] The key benefits of regular physical activity include reducing the risk of developing or dying from heart disease, diabetes, and high blood pressure, and helping to control weight and promote psychological well being.[8]

As for seniors, there is promising evidence that strength training and other forms of exercise in older adults preserve the ability to maintain independent living status, and to reduce the risk of falling. Physical activity also appears to relieve symptoms of depression and anxiety and improve mood.[9]

People realize the importance of exercise so they jog… for fifteen minutes. But studies show that power walking, with comfortable footwear, for thirty minutes is far more beneficial, and avoids the many complications that impact the body when jogging. Time is a premium for all of us today, so we need to use it as efficiently as possible. If a job has to be repeated because of undue attention the first time, it wastes time as well as resources.

So it is with healing. We can only expect positive results in proportion to how diligent and persistent we are in applying the true remedies as opposed to quick-fix drugs. But, like children, we are often too impatient for the finished product, and end up with something much less.

It was waffle day. Our four-year-old son Nathan was in his high chair and I was ready for the first waffle so I could eat and get off to work. The timer went off and out came a beautiful golden brown waffle, ready for fruit sauce and fruit topping. In went the batter for another waffle. Within a couple of minutes Nathan asked for his waffle. His mother told him to be patient. As soon as the timer went off, he would get his. Without missing a beat he responded, "Give me the timer".

As we reach for the simple agents of nature, apply them with care and prayer, we may fully expect the God of heaven to hear our heartfelt petitions and bestow the blessing we have requested of Him. But if we are not willing to work trustingly and patiently, well then, like Nathan all we are liable to get is soggy waffles.

Dr. Exercise recommends a minimum of twenty consecutive minutes of deep-breathing exercise every day. Whether around an outdoor track, back and forth in an indoor mall, doing several laps in the pool, down a wooded path, out along a sandy shore, on a bike or in a canoe, your body will love you. If you do not have a daily program, or your health is seriously compromised, start with a graduated program. Some seniors set themselves fifty paces, a city block, or the distance between power poles, and then add ten paces each day thereafter. Comfortable shoes can turn a chore into a joy. Head erect, shoulders back, breathing the crisp morning air, can there be anything finer?

Dr. Water

In a random survey in 2000, 2,818 American adults in fourteen metropolitan areas were asked how much water they drank each day. Only 34% claimed they drank eight or more eight-ounce servings daily (the recommended amount), while 28% only drank three or fewer servings, and nearly 10% said they didn't drink water at all. Coupled with the fact that the average American consumes 5.9 servings of dehydrating beverages (coffee, tea, caffeinated colas, and alcohol) daily, we see that upwards of 66% of the general public are chronically dehydrated.[10]

A 2000 article in *Women's World*, put the national dehydration average as high as 75%. The article went on to say even *mild* dehydration will slow down one's metabolism as much as 3%. Lack of water is the number one trigger of daytime fatigue. Preliminary research indicates that eight to ten glasses of water a day could significantly ease back and joint pain for up to 80% of sufferers. A mere 2% drop in body water can trigger fuzzy short-term memory, difficulty focusing on the computer screen or on a printed page, and trouble with basic math. According to the article, some research has shown that drinking five glasses of water daily decreases the risk of colon cancer by 45%, plus it can slash the risk of breast cancer by 79%, and, one is 50% less likely to develop bladder cancer. If these figures are even half correct, there is ample evidence to show that increasing one's water consumption is a good preventive measure for cancer.

Are you trying to lose weight? One glass of water before going to bed shut down midnight hunger pangs for almost 100% of the dieters in a University of

Washington study. In 37% of Americans, the thirst mechanism is so weak that a craving for water is often mistaken for hunger.[11] Are you drinking the amount of water you should every day?

Dehydration is not just a problem in America. It is also a worldwide phenomenon. Some researchers feel the above figures of low daily water consumption likely apply to half the world population.[12] While working in Nepal, Janie, RN, saw dehydration on a daily basis. The people knew that the contaminated water in the streams and rivers could be lethal, so they avoided it. Instead, people drank other beverages, or looked to relieve the symptoms that develop from not drinking adequate water by taking some quick-fix medicine, in our case the coveted American drugs.

Janie was in the clinic one day when a woman from a nearby community came with her elderly mother. After examining the mother, Janie concluded that she was severely dehydrated. She was of the lowest cast and poor. Her family was little able to care for her and she had no other recourse for help.

Janie asked to see the medications she was taking to relieve some vague symptoms. Through the interpreter, Janie asked if she could buy back her drugs, and give her a more powerful medicine. The woman gladly agreed.

Then Janie proceeded to explain that the woman must drink eight or more glasses of water a day, and she would soon be feeling much better. The woman agreed, but then asked for the strong medicine. Even after the third attempt, the interpreter was hardly able to convince the woman that Janie was referring to water. She and her daughter left discouraged and very disenchanted.

The next day Janie took a special hike down to the village to watch the woman drink her water. She did this every day for over a week. She was rewarded by seeing the elderly woman march up the hill one day with a large smile, and many thanks to the very wise American nurse.

According to Dr. Water, we would all be wise not to wait until we are old and sick before we "take the water of life freely"[13].

Dr. Sunshine

Sunlight, as you probably know, is the main factor in the production of Vitamin D. Both sunshine and Vitamin D affect the pineal and pituitary glands, which in turn affect the thyroid gland, which regulates all the body's processes, its use of fatty acids, proteins, and carbohydrates. Without sunshine, physical growth, building, and repair would be impossible.

Sunshine is also important in combating the mental illness known as Seasonal Adjustment Depression (SAD), which is directly related to how much light is available. Twenty minutes of exposure to intense incandescent light in the morning has changed SAD people into "normally" behaved people in a short period of time.

Dr. Paul Goldberg, a Cambridge researcher, found that neurological disease doesn't exist at the equator (*International Journal of Environmental Studies*). There, with all the abundant sunshine, the body produces two to four thousand IUs of Vitamin D per day compared to the Recommended Dietary Allowance of four hundred IUs.

Extra sunshine also seems to reduce the risk of multiple sclerosis (MS) for those people living at high altitudes and closer to the equator. High altitude above a thousand feet intensifies ultra violet exposure and reduces MS risk. Multiple sclerosis begins to develop as human populations move away from the most intense sunlight. Orthodox medicine has virtually overlooked this phenomenon in applying a remedy for MS.

On the Scottish islands of Shetland and Orkney, the cases of MS are among the highest in the world, much higher than the average one in a thousand seen in northern latitudes or smog polluted cities. Denmark's Faeroe Islands, two hundred miles northwest of the Shetlands, with exactly the same sunlight, dampness, overcast, and gloom, has less than the normal incidence of MS. What accounts for the difference? On the Faeroe Islands, seafood, high in Vitamin D, is abundant; on the Scottish isles, seafood is virtually non-existent in the diet. In Norwegian fishing communities where even margarine has a high fish oil content, MS rates are also low, but in Norwegian agricultural communities MS is high. The highest rates of all are in communities that have a high oat consumption, or high phytate consumption in addition to northern latitudes. Phytates bind with calcium and are antagonistic to Vitamin D. In Northern Scotland, where the MS rates are the highest in the world, high oat consumption, combined with a low seafood diet and lack of sunshine, has resulted in a perfect environment for MS.[14]

Sunshine is an extremely interesting subject and Dr. Sunshine highly recommends everyone get outside and study it on a regular basis. Treat the body to an hour or more of natural outdoor light even if it means putting on adequate sunscreen, head covering, or bundling up with mittens and scarves.

Dr. Temperance

The word temperance can mean the total abstinence of some hurtful product, or exercising restraint as far as good things are concerned. Alcohol, tobacco, caffeine products, and drugs are all pharmacologically classed as poisons, some more poisonous, some less. Temperance in regards to this class of products means total abstinence, while the exception to the rule would be the rare use as a medicinal. For example, caffeine drinks should never be used as a beverage, but because of their anti-spasmodic effects, they could be used on those rare occasions for seasickness or severe nausea. While water depravation is a major problem, we have an emerging generation of caffeine junkies who are unable to function without a jolt to jumpstart them in the morning, not just from coffee or tea but also from more and more chocolate and soft drinks.

As for Coke, it isn't just urban legend that phosphoric acid, the active ingredient in Coke, contributes to loss of bone mineral density leading to osteoporosis.[15] Studies show that regular soft drink users have a lower intake of calcium, magnesium, Vitamins A, C, and B_2. Nutritionists warn that soft drinks pose a major developmental risk for children when those beverages compete with a balanced diet. And, because poorly metabolized high fructose corn syrup is the sweetener, soft drinks are major contributors to the ballooning number of tubby teenagers predisposed to diabetes. No worries about fat calories with water. As for Coke, why not leave it to those farmers in India who, because it's cheaper than commercial poisons and doesn't need to be diluted, use Coke as an insecticide?

Temperance also includes self-control. That means, not too much of a good thing, not too much good food, or exercise, or sunshine, or rest, or water. Of course employing all eight NEWSTART® doctors at the same time works synergistically. That is, each part empowers the other parts, such that, working together, the net result is more than just the sum total of the parts. For example, if you're having trouble saying "NO" to some temptation, proper exercise and enough rest working together, can help to strengthen your will power. Once the willpower is energized you will find it easier to resist that last piece of…

Dr. Temperance is the more phlegmatic of the eight doctors and as such he extols the virtues of thinking things through, counting the costs, and keeping a cool head.

Dr. Air

Air is heaven's food for your lungs. Household air gets stale very quickly without good ventilation. Consequently that devitalised air requires more fuel to heat it in the winter. The same principle works for the body. Fresh, electrically charged air promotes physical and mental activity, and healing.

The vital ingredient in air is oxygen. While the whole body requires oxygen to live, it is the brain that fails first when there is not enough. We get our oxygen supply from the air we breathe. We get the best quality of air out of doors away from all forms of air pollution. We get the best delivery of oxygen when we are exerting ourselves physically and breathing deeply. Most folks breathe by filling the top of their lungs and end up using only a fraction of what is available. This form of breathing is a learned behaviour and not natural.

To understand how one should breathe properly watch a baby's stomach as they are lying flat on their backs. Their tummies will rise and fall as they breathe. The diaphragm, positioned below the lungs, pulls downward drawing air into the bottom of the lungs first. You can practice proper breathing by lying on your back with your hands on your stomach. Consciously pull your diaphragm down as you breathe in. Like the action of a bellows, it will fill the bottom of your lungs first, and your hands will rise. As you force your diaphragm to contract when you breathe out, your hands will fall. As you master deep breathing both brain and body will benefit with more oxygen even when you are not exercising. But, by exercising in any one of the ways mentioned earlier, your lungs will get all the "food" they need plus lots of extra oxygen to share with the rest of the body.

Do you live where you have poor air? For that matter do you live where you have poor water, poor light? How valuable is your health? Have you considered moving? Are you making plans and saving pennies? Big plans begin with little strategies and big investments begin with little savings. It is not a lack of money that keeps many who are financially able from making a home in the country for themselves and their children. And, it was not the abundance of money that prompted many, whom society counts poor, from establishing themselves outside the noise, the traffic, the pollution, and the many other growing problems associated with city life.

To live in the country would be very beneficial especially to children. An active, out-of-door life would develop health of both mind and body. Irrespective of what society counts as prosperous, fathers and mothers who possess a piece of land and a comfortable home, no matter how humble, may count themselves kings and queens.

For those living in the country, for those planning and saving to make the transition to a more rural setting, and for those who are not, all alike need to take regular opportunities to go out into nature and enjoy the life-giving currents of air. Whether along a warm pebble beach, a crisp mountain trail, a quiet meadow path, or a meandering singing stream, body, mind and spirit will be electrified and recharged by the natural stimulation of the entire person. That's Dr. Air's prescription for strength and joy and peace.

Dr. Rest

"Take rest; a field that has rested gives a bountiful crop", wrote the Roman poet Ovid (43 B.C. – A.D. 17). Everything needs rest, and people definitely do better with daily, weekly, and yearly periods of rest. As it was shown earlier, seven to eight hours of sleep a night helps to prolong life and promote health. But don't expect to slide into bed like you are running to home base and think you will fall asleep. The mind and body need to wind down their circuits. Shut off all stimulation, turn on soothing music, go for a leisurely walk, rock in the rocker in front of the fireplace or out on the porch, or take a warm bath. Some herbal tea, a good book, candle light, all these and more can help the weary body and mind prepare for the rejuvenating power of sleep.

Sleep is vital for body building and repairing. While 90% of a young child's growth happens when he's sleeping, for an adult's body, 90% of the healing and repairing happens as she's sleeping. But don't think you can overindulge on food or high calorie snacks and you will get the sleep you need. Remember, your whole digestive system is one long tube of muscles. If those muscles are still working, trying to digest food long after you have gone to bed, you will not wake up refreshed as you could have otherwise.

Having satisfied the body's need for physical rest there are still the needs of the mind and spirit. The fevered pursuit of money and things has eclipsed the real joys of life. It is more than just interesting trivia that in well over one hundred ancient and modern languages, the same day of the week carries the clear message of rest. From German to Chinese, from Russian to Arabian, from Hebrew to Tibetan, from Malayan to Swahili, from Latin to Greek, from Lithuanian to Hindi, around the world, ancient and modern idioms number the seventh-day Sabbath as a day of rest from otherwise secular pursuits. A lonely few languages, including modern English, have not only substituted the name, but also ignored man's inherent need for regular physical/mental rest from an otherwise stressful week.

Dr. Rest urges all to take time with loved ones, time out in nature, time alone meditating on the big questions of life, time exploring some hidden talent, time learning some art, time helping someone less privileged, time visiting the sick, time writing a letter. All these and more long for our quiet attention. Take rest.

Dr. Trust in Divine Power

We place a good deal of trust in that which is scientifically measurable. But what about those dimensions that are not measurable—the emotional and the spiritual? Today there are many self-help programs for any number of physical or psychological problems. Many of them are patterned after the famous twelve-step program of Alcoholics Anonymous, which has helped countless thousands back to a healthier lifestyle.

One of the central pillars of the program is the spiritual dimension—trust in Divine help. No matter what the background, no matter what the religion or faith, no matter what the circumstances, education or financial status, each new participant is invited to look to a higher power, outside the lonely self, to help them in the battle to reclaim their life. Dr. W. W. Bauer, former director for health education for the American Medical Association, explains the importance of our spiritual dimension this way: "Total health involves the spiritual, emotional, and social aspects of life as well as the physical."

There is a new trend in our secularized world that is not new at all. It is prayer. Among the indicators is a study led by David Eisenberg, MD, reporting that 35% of respondents to a 1997 telephone survey said they had used prayer for health-related problems that year. Also 99% of 296 physicians surveyed at a 1996 meeting of the American Academy of Family Physicians said religious beliefs can heal, while 75% said that others' prayers can promote healing. And today 60 of 126 medical schools in the U.S. offer courses on religion and spirituality.

In our post-modern society there is a growing disillusionment with the magic of science. Technology, for all its wonders and innovations, has not ushered in its prophetic age of world prosperity free from disease. Instead there is a growing interest in the paranormal—that which cannot be so nicely quantified and packaged. Prayer is one of those non-measureables, or so it has been thought to be.

Prayer

Prayer has, in recent years, come out of its closet into the public domain. In 1988 the first research article into the relation between prayer and healing was published in the *Southern Medical Journal.*[16] In his randomized, double blind study, cardiologist Randolph Byrd, MD, led the way for investigators intent upon learning whether distant prayer can help heal. Of 393 patients at the San Francisco General Hospital's Coronary Care Unit (CCU), half (192) were prayed for, while the others (201) were not. Dr. Byrd found that the prayed-for patients were less prone to congestive heart failure and cardiac arrest, and fewer of them needed diuretics and antibiotics than the control-group of patients who were not prayed for. Of those for whom prayer was offered, only three required antibiotics, compared with sixteen in the control group, a five-fold difference. Six of those who were prayed for suffered complications of the lung. This is compared with eighteen of those for whom no prayers were offered. A three-fold difference! Twelve of those who did not have the apparent advantage of prayers, required unusual methods of life support (intubations of the larynx and trachea).

Other studies, including at the Mayo Clinic, Harvard Univ., Duke Univ., and the Mid America Heart Institute, have since followed Dr. Byrd with more research on prayer with cardiac patients. William Harris, MD, of the Mid America Heart Institute in Kansas City, Missouri, followed Dr. Byrd with a study of 990 CCU patients, published in 1999. In all the studies, patients randomly assigned to the prayer groups appeared to fare better than those in the control group who were not prayed for. Cardiologist Mitchell Krucoff, of Duke University Medical Center, performs several cardiac catheterizations and angioplasties every day, and before each procedure, he prays.

Can consciously thinking of others in a supportive way really produce measurable results? Apparently so. It is an age-old truth that we can all read in the Holy Scriptures: "Pray for one another, that you may be healed. The effectual, fervent prayer of a righteous man helps more than you imagine."[17] As for the personal benefits of prayer, Dr. Trust recommends, "Come unto Me, all ye who labor and are heavy laden, and I will give you rest ... and ye shall find rest unto your souls."[18]

The eight-doctor NEWSTART® team, as we can see, is involved in more than just emergency palliative care. Its primary focus is maintenance and prevention. This brief chapter is just an introduction into what should be a study of a lifetime. Because we are dynamic beings living in a dynamic world, what serves us very well today does not mean it will work as well tomorrow. We are constantly changing. Our bodies go through cycles, they age, we may change where we live, the seasons change, our employment may change, accidents happen, all these and more fluctuations in our lives mean that, while we maintain basic rules of health, we will have to adjust them from time to time.

Where do you begin? Any one of the eight doctors is available for counseling. Resources abound. Remember, if health is a lifetime study, then be pragmatic and realistic in your approach. If, after experimenting with some program for a reasonable period of time, you do not get results, be prepared to try something else. Take bite size portions of information, and chew well before swallowing and taking your next bite.

If your circumstances demand some practical guidance and support, or if you feel a short retreat would be helpful, I have included a short list of lifestyle centers (American and International) that I am familiar with at www.charcoalremedies.com.

NEWSTART®

What many are looking for today is a new start, a new beginning for their body, their mind and their heart. The Weimar Institute of Health and Education in northern California, which pioneered the NEWSTART® program, specializes in treating coronary heart disease, diabetes, obesity, and hypertension. When I called Weimar, I asked if I could talk to someone who could tell me if they used charcoal at all in their program. Dr. Ing, MD, was kind enough to return my call. He is the medical director for Weimar Institute, and so I asked if they ever had occasion to use charcoal.

People attend their programs for any number of different health problems, and he reported that they do use charcoal for those who complain of diarrhea. He also said:

> "We use charcoal for infections and boils. We show our guests how to make poultices and explain that if they don't have some nice muslin cloth,

> they can always use a paper towel. Simply spread the paste over one half of the towel and fold the other half over. Then place the poultice over the infection, making sure there is good moist contact with the skin, cover that with some plastic and secure it with some tape."
>
> After pausing he added:
>
> "We also give a slurry drink of activated charcoal for patients with irritable bowel disease (IBD) and Crohn's. In fact it is so effective that when people call and are exploring the idea of coming to Weimar, I hesitate telling them over the phone that charcoal will help their symptoms clear up. I know if they do take charcoal they may end up not coming, and then lose the benefit of the whole NEWSTART® program."

Absolutely. Health is a matter of the whole person. Charcoal definitely does help the symptoms of IBD and Crohn's disease, but does nothing for the cause. The cause is a lifestyle issue, and that means education, equipping individuals with information including nutrition and healthful recipes, and then encouraging individuals to think and research for themselves.

As for his personal use, Dr. Ing mentioned that when they were working overseas, their teenage son got a nasty cut on his foot from the coral on the beach:

> "It became infected and my wife noticed a red streak traveling up his leg. She applied a charcoal poultice and by the next morning it was gone. Of course we use it for the occasional diarrhea too. My wife is more apt to use it than I am, and she feels it is very beneficial. For example she will place a tablet in her mouth against a canker sore, and she finds it heals more quickly."

It does us little good to educate others about the eight doctors if we do not put into practice these health principles in our own homes.

Education

Positive health education should begin in the home. When parents take the lead as primary health educators and role models, their children are less likely to be confused by artificial and conflicting information both in the media and in schools. As children see their parents being responsible stewards of their health, the children will follow. As they watch their parents apply the simple natural remedies available around them, they are not only intrigued but will be ready to accept them when they get sick. Familiarity on the part of the parent with natural remedies will naturally quiet the child's fears. This applies for charcoal as well. If parents begin when infants are small, giving charcoal orally for upset stomachs, and applying charcoal with a band-aid to the small insect bites, they will soon find charcoal is well accepted by children. Should some emergency arise, the children will not be intimidated if a larger dosage is offered or if a poultice is applied. However the old saying is still true, "An ounce of prevention is worth a pound of cure". In teaching total health, the emphasis should always be on preventing rather than curing. Only in recent years has this old philosophy of disease prevention gained renewed vigor.

Unfortunately, many of those who should be at the forefront of teaching the more enlightened principles of disease prevention, of whom many are employed in the healthcare system to care for the sick, are themselves in poor health, enslaved to drugs, and emotionally drained. Up until recent policy changes, friends and relatives coming to visit their sick loved ones at some hospital or nursing home, often had to walk past a cloud of blue haze as patients, nurses, and aids congregated outside to smoke. Coffee rooms and waiting rooms in hospitals continue to be supplied with food dispensers offering a poor quality of nutrition.

Without the constant stimulation of coffee, tobacco, or other drugs, many healthcare workers would simply not be able to function. Just as many of these individuals would think twice about taking their prized car to a mechanic who cares little about what kind of oil or gas he puts in his own car, just so the informed public is becoming more discerning about who they are willing to let manage their health concerns. The following ancient stanza sadly describes the reality that many healthcare workers are faced with, as they continue to emphasize the use of drug therapy:

> "What ails the physician that he dies of the disease that he would have cured in time gone by? There died alike he who administered the drug and he who took it, and he who imported and sold the drug, and he who bought it."[19]

Many healthcare workers have become martyrs for a system of medicine that not only failed to help their patients but also failed them. Drug therapy, as it is commonly practiced today, is not able to manage the prolific increase in disease. In fact, orthodox medicine, with its emphasis on treating disease with drugs rather than prevention, is being implicated more and more in the increase of disease. Until policy makers are more informed, strong enough and vigilant enough to pass and enforce laws to protect the sick and vulnerable, until the public demands that they will have something more than drugs, there will be no appreciable change. Clearly, if there is to be a change for the better, it will be a grass roots reform. It is left to the individual to take charge of personal health and that from the perspective of prevention. In the meantime, as in a more unenlightened era, some health reformers will continue to advocate common sense in the face of all manner of opposition.

Prevention

In 1849, Lord Palmerston, prime minister of England, was petitioned by the Scottish clergy to appoint a day of fasting and prayer to avert the cholera. He replied, in effect: "Cleanse and disinfect your streets and houses, promote cleanliness and health among the poor, and see that they are plentifully supplied with good food and raiment, and employ right sanitary measures generally, and you will have no occasion to fast and pray. Nor will the Lord hear your prayers while these, His preventives, remain unheeded."

The practical solution is to acquaint ourselves with the eight NEWSTART® doctors of health. There is no need to entertain feelings of helplessness when these tireless health professionals stand ready to take our case. One by one, as we employ these fundamental health principles to work on our behalf, we can expect to see a change for the better, not only in our own health, but also in the health of our children and loved ones.

On the other hand, the following verse parodies how nearsighted society becomes when its primary focus is treating instead of preventing disease.

A Fence or An Ambulance[20]

T'was a dangerous cliff as they freely confessed,
Though to walk near its edge was so pleasant,
But over its edge had slipped a Duke,
And full many a peasant.
So the people said something would have to be done,
But their projects did not at all tally.
Some said, "Put a fence around the edge of the cliff,"
Others, "An ambulance down in the valley."

The lament of the crowd was profound and loud,
As their hearts overflowed with their pity;
But the ambulance carried the cry of the day,
As it spread to the neighboring cities,
So a collection was made to accumulate aid,
And dwellers in highway and alley,
Gave dollars and cents not to furnish a fence,
But an ambulance down in the valley.

A Family Practice

"For the cliff is all right if you're careful", they said,
"And if folks ever slip and are falling;
It's not the slipping and falling that hurts them so much,
As the shock down below when they're stopping."
And so for the years as these mishaps occurred,
Quick forth would the rescuers sally,
To pick up the victims who fell from the cliff,
With the ambulance down in the valley.

Said one in his plea, "It's a marvel to me
That you'd give so much greater attention,
To repairing results than to curing the cause;
Why you'd much better aim at prevention.
For the mischief of course, should be stopped at its source;
Come friends and neighbors let us rally!
It makes far better sense to rely on a fence,
Than an ambulance down in the valley."

"He's wrong in his head," the majority said.
"He would end all our earnest endeavors.
He's the kind of a man that would shirk his responsible work,
But we will support it forever.
Aren't we picking up all just as fast as they fall,
And giving them care liberally?
Why, a superfluous fence is of no consequence,
If the ambulance works in the valley."

Now this story seems queer, as I've given it here,
But things oft occur which are stranger.
More humane we assert to repair the hurt,
Than the plan of removing the danger.
The best possible course would be to safeguard the source,
And to attend to things rationally.
Yes, build up the fence and let us dispense,
With the ambulance down in the valley.

—Joseph Malins (1895)

CR

[1] White, Ellen G, *2 Selected Messages*, Review & Herald, p. 298
[2] White, Ellen G, *The Ministry of Healing*, Pacific Press Publishing Association, p.127, 1942
[3] Donaldson, Stewart I, The Seven Health Practices, Well-Being, and Performance at Work: Evidence for the Value of Reaching Small and Underserved Worksites, *Preventive Medicine*, 24, 270-277, 1995
[4] Leading the group of healthcare workers is psychiatrists (7% of total deaths), then white female doctors (3.6% compared to .5% for women on average) and then white male doctors.
National Center for Health Statistics, Table 1-27, "Deaths from 282 Selected Causes, by 5-Year Age Groups, Race, and Sex: United States -1990", *Vital Statistics of the United States*, 1990
Frank et al., National Occupational Mortality Surveillance database, reported in "Mortality Rates and Causes Among U.S. Physicians," *American Journal of Preventive Medicine*, Vol. 19, No. 3, 2000
[5] Kassirer, JP, Doctor Discontent, *New England Journal of Medicine*, 339:1543-1544, 1998
[6] Physician's comment, National Public Radio, 11/11/95
[7] Shephard, Roy J, Exercise and Relaxation in Health Promotion, *Sports Medicine, Vol. 23, No. 4, 211*-216, April, 1997
[8] Pitts, Edward H, The Surgeon General's Call to Action, *Fitness Management, Vol. 12*, No. 9, 36-38, August 1996
[9] The Effects of Physical Activity on Health and Disease, Center for Disease Control, 85-172, 1996
[10] www.bottledwater.org/public/InfoForRepNatFactSheettest.htm
[11] *Women's World*, July, 2000
[12] Batmanghelidj, Fereydoon, *Your Body's Many Cries for Water.* Global Health Solutions, 1995
[13] Revelation 22:17
[14] Cooter, Stephan, PhD, *Sunshine and Health*, 1994
[15] Tucker, Katherine L, PhD. Cola Soft Drinks May Contribute to Lower Bone Mineral Density in Women, 25th Annual Meeting of the American Society for Bone and Mineral Research, Sept. 2003
[16] Byrd, Randolph C. Positive therapeutic effects of intercessory prayer in a coronary care population, *Southern Medical Journal*, 81, pp. 826-829, 1988
[17] James 5:16 – author's paraphrase
[18] Matthew 11:28, 29
[19] Verses upon the death in Baghdad of the physician Yuhanna ibn Masawayh in the year A.D. 857
[20] Joseph Malins, *The Best Loved Poems of the American People*, Garden City Publishing Company, Garden City, New York, 1936

Chapter 20 "First, Do No Harm"

"You medical people will have more lives to answer for in the other world than even we generals." —Napoleon Bonaparte (1769-1821)

Iatrogenic death—death "induced by a physician's words or therapy (used especially of a complication resulting from treatment)."[1]

Why should you, the reader, seriously consider using charcoal, or some other simple rational remedy, in treating any number of common ailments, rather than depending exclusively on conventional medicine, including drug therapy? Why should you think about "building a fence along the cliff" if there are thousands of "ambulances down in the valley"?

More and more, the medical system is falling into disrepute. In a study reported in the July 2000 issue of the *Journal of the American Medical Association* (*JAMA*), by Dr. Barbara Starfield of the Johns Hopkins School of Hygiene and Public Health, the startling evidence showed there were 225,000 deaths in America due to iatrogenic causes! The most significant number of these unnecessary deaths, 106,000, were due to the negative effects of **properly** prescribed drugs making them the fourth leading cause of death in the U.S.[2]

Dr. Starfield cautioned that as startling as this research was, it only represented hospitalized patients. It did not include negative effects that are associated with disability or discomfort. And, these estimates of death due to error are lower than those in other published reports. Also, these figures do not include deaths in nursing homes, emergency rooms, or in doctor's offices. Nevertheless, 225,000 iatrogenic deaths per year constitute the third leading cause of death in the United States, after deaths from cancer and heart disease![3] On a strictly monetary basis, depending on which published study you use, this adds up to $58.5 to $100 billion in extra costs! Walter Cronkite, well-known news anchor, painted a dismal picture: "America's healthcare system is neither healthy, caring, nor a system."[4] (1990)

Health Care Spending

In the year 2003 the U.S. healthcare spending reached $1.6 trillion, which represents 14% of the nation's gross national product. With the upsurge of interest in alternative medicine, pharmaceutical companies have lobbied hard to have lawmakers impose onerous restrictions on the neutroceutical industry, which produces dietary supplements. Simple remedies, by association, also came under the bombardment of slanderous campaigns attempting to undermine the value of nature's more benign remedies. Clearly the stakes are high.

The propaganda against natural medicine only succeeded in provoking an independent review of government-approved medicine. The research compared thousands of published, peer-reviewed scientific studies.[5] The numbers generated were higher than those cited in the *JAMA* article, but the conclusion was the same: there is something dreadfully wrong with American medicine. Too often it does more harm than good. Compelling evidence from this study unveiled astounding statistics on the 783,936 iatrogenic deaths per year that have resulted from conventional medicine at the staggering cost of $282 billion. Depending on which of the above studies one finds more credible, it still comes down to between two and six jumbo jets falling out of the sky each and every day! Is this phenomena limited to the U.S.?

Great Britain

According to a research team from University College London, almost 70,000 British patients die each year partly as a result of "adverse events" they suffer during hospital stays. One in ten patients admitted to hospital is harmed by complications, half caused by medical mistakes of some kind. The extra hospital days cost at least 1.5 billion dollars a year.[6] It should not be surprising that these research statistics continue to grow each year, since there have been no significant policy changes in the system. Deaths in England due to medical errors have risen 500% from 1990 to 2002. In a country notorious for gambling, it is still surprising to see policy makers willing to gamble at such high risks.

Australia

In a response to the *British Medical Journal*[7], Ron Law, a member of the New Zealand Ministry of Health Working Group advising on medical error, offered some enlightening information on deaths caused by drugs and medical errors. Official Australian government reports reveal that preventable medical error in hospitals is responsible for 11% of all deaths in Australia, which is about one of every nine deaths. If deaths from properly researched, properly registered, properly prescribed and properly used drugs were added along with preventable deaths due to private practice, it comes to a staggering 19%, which is almost one of every five deaths.

Canada

The newest study in Canada[8] (2004) found one in every thirteen patients treated in hospital is at risk of suffering an unintended injury or complication that results in death, disability or delayed hospital discharge. Of the approximately 2.5 million people who are admitted into hospital every year, about 9000 to 24,000 people die as a result of preventable medical errors. As for costs, it is estimated that preventable hospital-born infections alone, cost an estimated $100 million annually.[9] Experts say that the totals, when taking into account population size and the numbers of patients in hospitals, are not that different from preventable medical error percentages and subsequent deaths in the U.S., Great Britain and New Zealand. So which drugs are the most deadly?

Worse Drug Offenders

The leading causes of adverse drug reactions are antibiotics (17%), cardiovascular drugs (17%), chemotherapy (15%), and analgesics and anti-inflammatory agents (15%).[10]

Antibiotics

Dr. Richard Besser, of the Center for Disease Control and Prevention (CDC), in 2003, said the number of unnecessary antibiotics prescribed annually for viral infections was in the tens of millions.[11, 12] The CDC posts this public awareness on its website: "Are you aware that colds, flu, and most sore throats and bronchitis are caused by viruses? Did you know that antibiotics do not help fight viruses? It's true. Plus, taking antibiotics when you have a virus may do more harm than good. Taking antibiotics when they are not needed increases your risk of getting an infection later that resists antibiotic treatment."[13]

In America, over three million pounds of antibiotics are used every year on humans. This amount is enough to give every man, woman and child ten teaspoons of pure antibiotics per year. The CDC promotes programs to educate not only the general public about inappropriate use of antibiotics, but also medical schools. Why the medical schools?

Drug Pushing Money

In 1981 the drug industry "gave" $292 million to colleges and universities. In 1991 it "gave" $2.1 billion.[14] This represented only 20% of the total $10 billion spent on sales and marketing, an estimated $6 billion of which was spent on promotions to individual physicians.[15] This breaks down to an estimated $8,000 per physician per year.[16]

In the same year, the drug giants only spent $9 billion on drug research and development.[17] Mind you, this is a paltry sum when you consider in 2002 the aggregate sales of drugs between the ten top drug companies was a cool $217 billion. This translates into an 18.5% profit for shareholders, easily outstripping all other Fortune 500 companies.

In order to reach the widest audience possible, drug companies are no longer just targeting medical doctors. A report in the *New England Journal of Medicine*[18] showed that by 1995 drug companies had tripled the amount of money allotted to direct advertising of prescription drugs to consumers. The majority of the money is spent on slick television ads. From 1996 to 2000, spending rose from $791 million to nearly $2.5 billion—15% of the total pharmaceutical advertising budget. The stakes are huge, but so are the penalties.

As an example, after nearly sixty years of drug companies and doctors promoting Hormone Replacement Therapy (HRT), in December, 2000, a government scientific advisory panel recommended that synthetic estrogen be added to the nation's list of cancer-causing agents. Results of the "Million Women Study" on HRT and breast cancer in England, published in *The Lancet*, August, 2003,[19] proposed that over the past decade, use of HRT by British women aged fifty to sixty-four has resulted in an extra 20,000 breast cancers. This study does not address breast cancer, stroke, uterine cancer, or heart disease due to HRT used by American women.

What about breast-cancer screening as a preventive? According to *The Lancet* (1995), the average annual cost per life "saved" by mammography is around $1.2 million.[20] Is it really more research that we need, as we are told over and over again?

"First, Do No harm"

Traditionally, the promise every doctor makes on graduation is, "First, do no harm." On top of that, both doctors and nurses pledge, "I will give no deadly/harmful drug."[21] But there have only been brief periods when, as a body, they can claim they have collectively honored their vows.

Strikes

It has been well noted that when doctors or nurses go on strike for an increase in money, there is a corresponding decrease in deaths.[22] There are several studies on this not-so-unexpected observation. Beginning March, 2000, "industrial action" (strikes) was begun by doctors in Israel, which clearly seemed to be good for their patients' health. According to a survey of burial societies, death rates dropped considerably in most of the country when physicians in public hospitals implemented a program of "work sanctions".[23] These results are in line with a similar 17% reduction of deaths during a doctor's strike in Los Angeles, California, in 1976[24], and a 35% drop in the same year in Bogotá, Columbia, and a 50% drop in Israel in 1973. Some will dispute these statistics, but it is clear in all the recorded strikes, when doctors or nurses refused to work, there were NO increases in deaths or illnesses.

Guns or Doctors

As quoted in the beginning of the chapter, Napoleon made a bold accusation against the medical fraternity in the early 1800s. How would that accusation stand up today? Here are some other numbers to ponder:

Year 2002

- Number of physicians in the US = 781,000
- Iatrogenic deaths per year = 225,000 to 783,936
- Accidental deaths per physician = 0.288 to 1.004

- Number of recreational gun owners in the US = 80,000,000
- Number of accidental gun deaths (all age groups) = 1,500
- Accidental deaths per gun owner = 0.0000188

We are accustomed to thinking of soldiers as killing in times of conflict, but these figures suggest doctors, as an organization, are approximately 15,300 to 53,400 times more likely to accidentally kill someone than recreational gun owners, every year.

On the other hand, Napoleon was, as far as his own personal health was concerned, ahead of his times. In place of drugs and drug peddlers, he studiously avoided "accidents" by prescribing to at least part of the NEWSTART® prevention philosophy. He wrote, "Water, air, and cleanness are the chief articles in my pharmacy."

Today we can read of the gross shortcomings of medicine in the days of Napoleon when, for example, hygiene was so blatantly ignored. Ignaz Semmelweis, who was so instrumental in promoting sound hygienic practices in the mid 1800s recalled, "There were heart rendering scenes when [pregnant] patients knelt down, wringing their hands, to beg for a transfer [to the midwifery division]" and not have to deliver in the new wing of the Viennese General Hospital (1846). Semmelweis noticed that three times as many women were dying at the hands of the medical students than at the hands of the midwifery students from puerperal fever, commonly known at the time as, "the black death of the childbed". Why? Because, after doing autopsies, the medical students proceeded on to assist in delivering babies, without washing their hands.

The *New England Journal of Medicine* reported 140 years later: "Hand washing is considered the single most important procedure in preventing nosocomial [hospital-borne] infections ... [but] compliance of healthcare workers with recommended hand washing practices remains unacceptably low ... We found that, on the average, hospital personnel washed their hands after contact with [intensive care unit] patients less than half the time. Physicians were among the worst offenders."[25] When a case of nosocomial infection does develop, physicians routinely turn to drugs and antibiotics to try and reverse the failure to follow fundamental rules of cleanliness. Incidentally, there are tens of thousands of "accidental" deaths each year that are attributed to nosocomial infections contracted in hospitals.[26]

Drug Deaths

While worlds apart on other issues (including guns), it would seem Napoleon and Ellen White agreed on some things. White wrote: "More deaths have been caused by drug taking than from all other causes combined. If there was in the land one physician in the place of thousands, a vast amount of premature mortality would be prevented. Multitudes of physicians and multitudes of drugs, have cursed the inhabitants of the earth, and carried thousands and tens of thousands to untimely graves."[27] (1864) We are often told by some, that the drugs of Napoleon's day and White's day, were such things as strychnine, arsenic, mercury, quinine, antimony, and the like, not to be confused with the benign drugs of today. "It was the quacks who used those drugs and not the docs", they say.

In fact, it was the allopathic doctors who promoted those deadly drugs the most passionately and vigorously. Homeopathic doctors had abandoned most poisonous drugs early in the century. In the 1870s, osteopaths and naturopaths likewise chose more rational treatments, such as nutrition, exercise and hydrotherapy, in place of mercury, arsenic, etc. It was these doctors, and especially women doctors who advocated a more holistic approach, who were branded as "quacks". But strychnine, arsenic, mercury, quinine, and their ilk are still in use today. They are regulated and the only ones allowed to prescribe them are licensed doctors. One is forced to ask, "Who are the quacks?" Sad to say, science has concocted even more powerful drugs than the doctors or quacks ever dreamed of in the 1800s. They leave people emaciated, bald, and nauseated without killing them outright.

Writing again to the world-famous doctor at Battle Creek, White counsels Dr. J. H. Kellogg:

> "… in no case are you to stand as do the physicians of the world to exalt allopathy above every other practice, and call all other methods quackery and error; for from the beginning to the present time the results of allopathy [using drugs as medicines to suppress symptoms] have made an objectionable showing … Drug medication has broken up the power of the human machinery, and the patients have died."[28]

White continues her unflinching assault on those men and women who, knowing full well the deadly effects of their drugs, ignore their conscience and continue to serve the sick their poisonous medicines:

> "There is a terrible account to be rendered to God by men who have so little regard for human life as to treat the body so ruthlessly in dealing out their drugs." Yet, as she adds, "We are not excusable if through ignorance we destroy God's building by taking into our stomachs poisonous drugs under a variety of names we do not understand. It is our duty to refuse all such prescriptions." [29]

Ultimately we are responsible for our own health. If we have been blessed with good minds, then it is inexcusable if we continue to load the body and mind down with drugs when we are sick.

Other lone voices, those "one in a thousand", have been willing to speak up. Long before White, Benjamin Franklin (1706–1790), famous American statesman and inventor, wrote, "He's the best physician that knows the worthlessness of the most medicines." Frank Billings, MD, was one of the founders and first Presidents of the American Medical Association (1903). "I place no confidence in drug therapeutics. Drugs, with the exception of two, are valueless as cures." Charles Mayo, MD, (1916), co-founder of the famous Mayo Clinic, Rochester, Minnesota, said, "The drugless healer is one of the best things that has come into the life of the present. The sooner patients can be removed from the depressing influence of general hospital life, the more rapid their convalescence."[30]

However, White reminded all who launch out to explore better methods, that they will meet opposition. "All who leave the common track of custom, and advocate reform, will be opposed, accounted mad, insane, radical, let them pursue ever so consistent a course."[31]

So it was with Dr. Semmelweis. The very next year after requiring every medical student to wash his hands with a chlorine solution before making an examination, the death rate at the Vienna Hospital plummeted. For the first time in its history, the mortality rate at the medical school fell below that of the school of midwives.

But instead of being knighted, he was summarily dismissed from the Royal Society of Medicine. So, he lectured, he wrote papers, and he continued to be ridiculed. He then turned from academics to polemics. He published open letters to midwifery professors. "Your teaching … is based on the dead bodies of … women slaughtered through ignorance. If … you continue to teach your students and midwives that puerperal fever is an ordinary epidemic disease, I proclaim you before God and the world to be an assassin."[32]

He took to the streets with his circulars vainly hoping to alarm the citizens. "The peril of childbed fever menaces your life! Beware of doctors for they will kill you … Unless everything that touches you is washed with soap and water and then chlorine solution, you will die and your child with you!"

As a consequence, Semmelweis, at the age of 47, the father of three young children, was committed to an insane asylum in Vienna. He attempted to escape, but was forcibly restrained by several guards, secured in a straight jacket, and confined in a darkened cell. Quoting from the *Bulletin of the History of Medicine,* "He was not in the asylum for long. Thirteen days after admission he was dead." From the autopsy report: "It is obvious that these horrible injuries were … the consequences of brutal beating, tying down, trampling underfoot."[33]

As Semmelweis protested reforms in Europe, just so did White advocate in America. "Those who make a practice of taking drugs sin against their intelligence and endanger their whole afterlife. There are herbs that are harmless, the use of which will tide over many apparently serious difficulties. But if all would seek to become intelligent in regard to their bodily necessities, sickness would be rare instead of common. An ounce of prevention is worth a pound of cure."[34]

When it comes to the poisonous effects of drugs, overdoses, adverse effects, and other drug–related complications, more and more people are discovering that an ounce of charcoal can go a long way in helping to remedy the results of pounds and pounds of drugs. Once someone realizes, that the drugs they are taking are doing them only harm, and look for something to help rid the system of these poisons, they will find charcoal a wonderful cleanser. If the health is not irreversibly damaged, and if the sick person is put on a well-regulated program like NEWSTART®, it will be found that charcoal can help the body to detoxify itself.

Unfortunately, many turn to some holistic approach to healing only after they have been given up on by orthodox medicine. Only then are some willing to try some benign remedy like charcoal. It is then that nature's humble doctors are expected to remedy, not only the original symptoms, but also the added complications of many drugs.

Sir William Osler, MD, Physician-in-chief to the John Hopkins Hospital, Baltimore, Maryland, recognized that drugs do not promote healing, but rather they prevent healing. "The person who takes medicine must recover twice, once from the disease and once from the medicine." Elmer Lee, MD, past Vice President, Academy of Medicine, emphatically agreed, "Medical practice has neither philosophy nor common sense to recommend it. In sickness the body is already loaded with impurities. By taking drug-medicines more impurities are added, thereby the case is further embarrassed and harder to cure." If the sufferer recovers then it is assumed it was due to the benefits of drugging. If they finally expire under the weight of abused nature, it is called a dispensation of God.

Hypocritic Oath?

The original Hippocratic oath begins with a pledge of allegiance to mythological gods, the priesthood of sorcerers, and their Art. As for treatment, it first promoted the importance of nutrition, while condemning drugging, euthanasia, abortion, and surgery. Yet it still endorsed vomiting, purging or bloodletting to free the system of disease. Obviously little remains of Hippocrates' original reforms, and while conventional medicine has continued on with the practice of vomiting and purging, it also favors blood thinning (blood-letters are now called phlebotomists and are licensed specialists), drugging, burning, and cutting. If anything, judging from the statistics mentioned earlier from the JAMA, modern health care systems today are more a curse than twenty-four centuries ago. Millions more die, and what was supposed to be a service to the community, not only has become big business, but also drains the world of trillions of dollars annually. So-called modern medicine has become a blight upon society and a continual embarrassment to conscientious doctors, nurses, and other healthcare providers.

People are increasingly distrustful with standard orthodox medicine. French dramatist Moliere, in his 1664 satire "A Physician in Spite of Himself", portrayed the general cynicism towards the business of drugging in his day. "I find medicine is the best of all trades because whether you do any good or not you still get your money." Over three hundred years later, people increasingly feel just as disenchanted towards modern medicine.

The *New York Times* article "As Doctors Write Prescriptions, Drug Company Writes a Check"[35] (June, 2004), begins: 'The check for $10,000 arrived in the mail unsolicited. The doctor who received it from the drug maker ["A"] said it was made out to him personally in exchange for an attached "consulting" agreement that required nothing other than his commitment to prescribe the company's medicines. Two other physicians said in separate interviews, that they too received checks unbidden from ["A"], one of the world's biggest drug companies. "I threw mine away," said the first doctor, who spoke on the condition of anonymity because of concern about being drawn into a federal inquiry into the matter."

We should be thankful there are still health practitioners and health researchers who are not willing to be bought and sold.

Which Way?

Reports such as Dr. Stanfield's, quoted at the beginning of this chapter, outlining the death toll of patients at doctor's hands, began to increase in the medical literature at the turn of the millennium. Since then, there has been virtually no change in the approach to medicine by the medical fraternity and no significant demand for it by the public. What incentive is there for change if the American Medical Association, and its journal, receive roughly fifty percent of its revenue from the pharmaceutical industry? It reminds me of a fanciful conversation between a little lost girl (standing at a junction of several roads) and a talking Cat.

Alice (all in a tizzy) "Would you tell me please which way I aught to go from here?"

"That depends a great deal on where you want to go to," said the cool Cat.

"I don't much care!" said Alice in a tearful whimpering voice.

"Then it doesn't much matter which way you go", said the Cat.

Indeed there remain many ways the so-called healthcare systems may choose to go, but the reality is there is little will for change. People still prefer to pop a magic pill to treat disease, rather than try and prevent disease. Whatever reforms do come about, will come through individual doctors, nurses, other health practitioners, researchers, and the ones they serve. As patients, if you and I show no inclination to become responsibly informed about our own bodies, and the nature of the remedies available, then we should not be surprised if we end up as just another statistic in some list of iatrogenic deaths.

Some Facts You Seldom Get to Read

Over a hundred years ago, Ellen White wrote of the greed of some licensed doctors, on the backs of the sick and the poor:

> "The exorbitant price charged by physicians … when called upon to attend suffering humanity is robbery, fraud. God gave physicians their wisdom and skill. It is not man who saves life; it is the Great Restorer. But poor men are often charged for services they never received."[36]

It is hard to imagine her words weren't penned yesterday.

According to an article in *Medical Economics* (1998), family practitioners who make more than $250,000 a year (after all expenses), do so because they see an average of 164 patients a week. If they see 150 patients per week, they average $178,000 a year. And if they see 50 a week, they only net $146,000.[37] But reducing people into mere numbers is not limited to doctors. One Health Maintenance Organization (HMO) executive told the *Wall Street Journal:*

> "We see people as numbers, not patients. It's easier to make a decision. Just like Ford, we're a mass production assembly line and there is no room for the human equation in the bottom line. Profits are king."[38]

A Better Way

It is often asked, "If charcoal rapidly and significantly lowers dangerous levels of cholesterol, if charcoal disinfects and cleanses wounds so effectively, if charcoal counteracts so many poisonous bites, if charcoal helps to rescue failing livers and kidneys, if … if charcoal does all these things as the research recorded in the many different medical journals testify, why is it not more widely used and publicized?" The answer is simple, "No money".

In the face of all this greed, should one whisper? Ella Wheeler Wilcox (1850-1919) wrote, "To sin by silence when we should protest makes cowards out of men".

Like Dr. Semmelweis, White did not mince her words when she warned about the dire effects of taking drugs:

> "The endless variety of medicines in the market, the numerous advertisements of NEW drugs and mixtures, all of which, as they say, do WONDERFUL cures, kill hundreds where they benefit one."[39]

In an address before the Massachusetts Medical Society, Oliver Wendell Holmes, MD, (1809-1894) gave the same diagnosis of the medicines of their day:

> "I believe that if the whole Materia Medica as now used, could be sunk to the bottom of the sea, it would be all the better for mankind, and all the worse for the fishes."[40]

"Haven't things really changed," you ask, "since that was penned over a hundred years ago? Aren't there fewer drugs being prescribed? Aren't they really less deadly? Don't they claim fewer lives?" Obviously not. But, what would people use if such a heroic effort were made to reform the health care systems—simple, inexpensive, benign natural remedies? We have looked at eight well-qualified time-tested doctors, and, in stark contrast to the giant drug cartels and their licensed drug dealers, may I reintroduce our humble protagonist.

In concluding his exhaustive book on the unsurpassed efficacy and affordability of charcoal as a home remedy for poisoning, David Cooney makes this sad observation: "Unless some profit motive exists for manufacturers and local pharmacies, the use of charcoal in the home *will never* come about."[41] On the other hand, a few pages earlier, the author also points out, "By now, the reader should be convinced that ever-new uses for the remarkable substance known as activated charcoal *will continue* to evolve."[42]

Knowledge is power, and personal experience cannot be contradicted. Like the lone butterfly moving across a blighted meadow, the impact of one's shared experience should not be undervalued. In spite of the studied indifference of the pharmaceuticals, pharmacists, and the medical fraternity, charcoal's reputation continues to spread. While drug companies continue to multiply new poisons under new names, one charcoal manufacturer happily proclaims: "Today a 1000 applications ... tomorrow a 1001."[43]

While doctors and nurses still recite promises, only charcoal follows through on its proven history to "first do no harm". CR

---

[1] WordNet® 2.0, © 2003 Princeton University

[2] Barbara Starfield, MD, *Journal American Medical Association* Vol. 284 July 26, 2000

[3] In the year 2001 the number of deaths attributable to heart disease was 699,697, while the number of deaths attributable to cancer was 553,251. (U.S. National Center for Health Statistics. National Vital Statistics Report, vol. 51, no. 5, March 14, 2003

[4] Cronkite, Walter, Borderline Medicine, PBS documentary, Dec.17, 1990

[5] Gary Null, PhD, Carolyn Dean, MD, ND, Martin Feldman, MD, Debora Rasio, MD, Dorothy Smith PhD., *Death by Medicine*, October, 2003

[6] *British Medical Journal*, 517; 548; 562, March 4, 2001

[7] *British Medical Journal*, 321: 1178A (emailed response), Nov. 11, 2000

[8] G. Ross Baker, Peter G. Norton, Virginia Flintoft, Régis Blais, Adalsteinn Brown, Jafna Cox, Ed Etchells, William A. Ghali, Philip Hébert, Sumit R. Majumdar, Maeve O'Beirne, Luz Palacios-Derflingher, Robert J Reid, Sam Sheps and Robyn Tamblyn, The Canadian Adverse Events Study, the incidence of adverse events among hospital patients in Canada, *CMAJ*, 170 (11), May 25, 2004

[9] CBC, The Current, March22, 2005

[10] Suh DC, Woodall BS, Shin SK, Hermes-De Santis ER. Clinical and economic impact of adverse drug reactions in hospitalized patients. *Annals of Pharmacotherapy*, 34(12):1373-9, Dec, 2000

[11] Rabin R, Caution About Overuse of Antibiotics. *Newsday*. Sept. 18, 2003

[12] http://www.cdc.gov/drugresistance/community/

[13] http://www.cdc.gov/drugresistance/community/snortsnifflesneezespot/index.htm

[14] Crossen C, Tainted Truth: *The Manipulation of Fact in America*, Touchstone Books, 1996

[15] "Pharmaceuticals, Inc." *PNHP Newsletter*, March 5, 1999

[16] Gibbons, RV, et al., A Comparison of Physicians' and Patients' Attitudes towards Pharmaceutical Industry Gifts, *Journal Of General Internal Medicine*, 13:151-154, 1998

[17] Sherrill, R, Medicine and the Madness of the Market, *Nation*, 44-71, January 9, 1995

[18] Rosenthal MB, Berndt ER, Donohue JM, Frank RG, Epstein AM. Promotion of prescription drugs to consumers. *New England Journal of Medicine*, 346(7):498-505, Feb 14, 2002
[19] Beral V; Million Women Study Collaborators. Breast cancer and hormone-replacement therapy in the Million Women Study. *The Lancet*, 362(9382):419-27, Aug 9, 2003
[20] Wright CJ and Mueller CB, Screening Mammography and Public Health Policy, *The Lancet*, 346:29-32, 1995
[21] Hippocratic Oath. It also states, "I will give no deadly medicine to any one if asked, nor suggest any such counsel." Similar to the Florence Nightingale Pledge: "I will not take or knowingly administer any harmful drug"
[22] Martin O'Malley & Owen Wood, CBC News Online, When doctors walk off the job http://www.cbc.ca/news/background/healthcare/doctors_strike.html
[23] Judy Siegel-Itzkovich, Doctors' strike in Israel may be good for health. *British Medical Journal*, June 10, 2000
[24] *American Journal of Public Health*, 69 (5):437-43, May, 1979
[25] Albert RK and Condie F, Hand-Washing Patterns in Medical Intensive-Care Units, *New England Journal of Medicine*, 304:1465-1466, 1981
[26] Inlander, CB, This Won't Hurt Allentown, People's Medical Society, 1998
[27] White, Ellen G, *Spiritual Gifts, Vol. 4*, p 133
[28] White, Ellen G, Letter 67, 1899
[29] White, Ellen G, *Healthful Living*, p 246
[30] Mayo, Charles, MD, *The Lancet*, 1916
[31] White, Ellen G, *Testimonies Vol. 2*, p. 377
[32] Elek SD, Semmelweis and the Oath of Hippocrates, *Proceedings of the Royal Society of Medicine* 59:346-352, 1966
[33] Carter, KS, S Abbott and JL Siebach, Five Documents Relating to the Final Illness and Death of Ignaz Semmelweis, *Bulletin of the History of Medicine*, 69:255-270, 1995
[34] White, Ellen G, *Selected Messages Book 2*, p 290
[35] Harris, Gardiner, "Medical Marketing – Treatment by Incentive; As Doctors Write Prescriptions, Drug Company Writes a Check". *New York Times*, June 27, 2004
[36] White, Ellen G, *Medical Ministry*, p 122
[37] Income Rises in Busier Practices and With Time Invested, *Medical Economics*, p. 181, 9/7 1998
[38] *Wall Street Journal*, 18 June 1997
[39] White, Ellen G, *Healthful Living*, page 245 (emphasis supplied)
[40] Oliver Wendell Holmes, MD, (1809-1894), poet and Harvard Medical School professor father of the great dissenting Supreme Court Justice Oliver Wendell Homes
[41] Cooney, David, O, *Activated Charcoal in Medical Applications*, Marcel Dekker, Inc., p.575, (emphasis supplied) 1995
[42] Ibid. p. 564 (emphasis supplied)
[43] NORIT motto

Chapter 21 None of These Diseases

Over 3,500 years ago, in the presence of the assembled dignitaries of Egypt, Moses, the shepherd prince, delivers God's message to Pharaoh, "Let My people go". The haughty monarch replies, "Who is God that I should let these slaves go free?" In answer, as he was commanded, Moses casts down his shepherd's rod and it becomes a serpent! Recovering from his amazement, Pharaoh challenges the God of heaven. As a god on earth, he summons his wise men who work their enchantments, casting down their rods, which immediately begin to slither across the floor like snakes! But, the shepherd's "serpent" moves in and easily swallows all the "serpents" of the sorcerers! In the weeks that follow, one plague after another falls upon the land of Egypt until Pharaoh's medicine men are forced to acknowledge that the power of Moses' plagues was "the finger of God" and not just magic. Nevertheless, Pharaoh obstinately refuses to acknowledge any higher power, until the Red Sea at last swallows him and his hordes.[1]

This fascinating ancient account surrounding the emancipation of the Hebrew nation from Egypt is recorded in the book of Exodus. It is within this same history that we see the development of a healthcare system that was to stand distinct from the practices of the day. It was God's plan to emancipate the Hebrews from more than just slave labor.

The Medicine of Egypt

One would think that 3,500 years ago the healthcare system of Egypt was so primitive it could in no way be compared to our modern practices. But, the ancient Egyptian knowledge of science in many ways eclipses even that of the 21st century. Along with their superior knowledge of embalming the dead, the Egyptian healthcare system had developed diagnostic and medical procedures that are modeled by modern-day protocols.

According to the Greek historian Herodotus, there was a high degree of specialization among Egyptian physicians: "The practice of medicine is very specialized among them. Each physician treats just one disease. The country is full of physicians, some treat the eye, some the teeth, some of what belongs to the abdomen, and others internal diseases."[2] There were brain surgeons, podiatrists, dentists, and pharmacists. As today, their medicines included benign herbs, poisonous plants, lead-based ointments, and the venom of scorpions and snakes.

However, medicine did not work alone. Magic was the main component of healing. Physical symptoms were relieved by physical medicines, but only magic could effect a cure. The doctors of sorcery were also the priests of magic who employed their drugs along with their invocations to the gods.

While the standard Egyptian food pyramid included whole grains, fruits and vegetables, it also encompassed foods that suited all the cravings of appetite. Though meat consumption was low compared to modern levels, there were no restrictions.

Among the ancient Egyptian culture, malnutrition was the norm even among the upper levels of society.[3] Historical records have shown that, while obesity was rare, and there was a general knowledge of the importance of nutrition, prevention, and self-control, disease was quite prevalent. Those diseases we count as common today were common then. From stomachaches to headaches, from the common cold to dysentery, measles, pneumonia, cancers, eye infections, arthritis, arteriosclerosis, dental cavities and dementia, all these and more plagued ancient Egypt as they do industrialized nations today.[4]

A Revolutionary New System

On the other hand, the Hebrews were handed down explicit dietary laws that, while focusing on a mainly vegetarian menu, only allowed for animal products from certain herbivores, while excluding all scavenger animals and carnivores. It strictly prohibited from the diet the blood and fat of animals, the two foods still plagued with health risks three thousand years later. Fermented foods and beverages were also avoided. Along with dietary laws, rules for sanitation and hygiene were carefully codified. Today, science recognizes these health laws as equal to if not better than any national healthcare program anywhere.

God promised the Hebrews that if they would follow His laws, including the code of health, they would be spared from the diseases of Egypt. God said, "I will put none of these diseases upon you."[5] They would become a beacon of health to surrounding nations who would come to learn of their advanced knowledge. But, slavery does not die easily.

While on their journey across the Sinai desert, the Hebrews began to complain of their circumstances, their leaders, and especially of their diet. They bitterly lamented leaving Egypt and its savory foods. Suddenly the very ground around them seemed to come alive with deadly snakes slithering everywhere. Thousands fell victim to the fatal snakebites. To stem the plague of serpents, Moses was commanded to make a brazen form of a dead snake and wrap it around a pole. The promise was made that when this dead snake was lifted up, if all who had been bitten would but look at the uplifted snake, their sins of ingratitude and complaining would be forgiven, the venom poisoning their life would be miraculously swallowed up, and they would recover.[6]

Recognizing man's proclivity to insubordination, God provided this additional remedy for those who presumed to ignore the laws of preventive health. When, through disobedience, some fell victim to disease, if they would but look in faith to a loving and merciful God, their acts of disobedience would be forgiven, they could once again resume their walk of obedience to God's laws, and expect healing.

The Counterfeit

That serpent, entwined about the pole, has since been taken by the medical and pharmaceutical world, with some modifications, as their symbol of healing. But, while they have claimed the emblem of healing by faith, they have, to a great extent, incorporated the methods and superstition employed by the priest-doctors of ancient Egypt.

Our modern-day words 'pharmaceutical', 'pharmacy', and 'pharmacist' are derived from the classical Greek word 'pharmakos,' which stood for poisoner, sorcerer, and magician.[7] It was used in that sense by the medicine men of Hippocrates' day. *Webster's 1986 New World Dictionary* takes note of this historical fact when it defines 'pharmaceutical' as "to practice witchcraft; sorcery". While the modern use of the word 'pharmacy' has slowly been redefined, the ancient craft of sorcery, with its drugs and incantations, finds its more sophisticated counterpart in modern day medicine, and the principle of mysticism is still the power it exercises over the general public.

At the beginning of the 20th century, Charles E. Page, MD, an outspoken American proponent of hygiene, as well as critic of certain prevailing medical practices, gave this evaluation of the increasing custom of drugging people: "The cause of most disease is in the poisonous drugs physicians superstitiously give in order to effect a cure." The intricate names given to poisonous drugs and therapies are used to cover up their true nature, so that none will know what is given them as remedies unless they obtain a dictionary to find out the meaning of these names. It is this sense of mystery that is attached to drugs, that is so reminiscent of ancient sorcerers.

Today, as in the days of ancient Egypt, there is a general knowledge of the importance of nutrition, prevention and self-control, but they are given only token lip service. Rather than disease prevention, the emphasis remains focused on treating disease with poisonous pharmaceuticals and expensive therapies. Modern medicine has not fundamentally changed from ancient Egyptian medicine, either in philosophy or in practice. Where some drug has benefited a few, hundreds if not thousands have died prematurely. The diseases common then are still common today, and if not for the major benefits of improved sanitation, labor-saving technologies, and food availability, average life spans would not be significantly improved over a hundred years ago.

Under Egyptian medicine, one could disobey the laws of health, treat the resulting symptoms of disease with drugs or magic, and then return to disobeying the same laws of health, presuming that drugs and magic would indefinitely come to the rescue. But, there are limits as to how much abuse nature will suffer, beyond which broken health will not naturally recover. And, there are boundaries to God's limitless mercy. The prevailing notion that one can "bungee jump through fire" and not get burned is as fatal a presumption today as it was so long ago.

The contest that began between Moses and the sorcerers of Egypt 3,500 years ago continues to be played out today between the modern day practice of drugging, and nature's simple remedies. While the giant pharmaceuticals have mounted a campaign to discredit their non-pharmaceutical competitors, the growing number of deaths attributed to properly prescribed drugs by doctors in hospitals, tells the informed person that they need to look for more benign alternatives. As long as pharmaceutical companies are the principal patrons of American and other national medical associations, there is an ethical compromise that puts the general public at constant risk. The average person must assume the role of gatekeeper for their health. They have a right and a duty to choose for themselves.

Medical Freedom

Those who won their religious and civil freedoms from the tyrannies of the old world realized they needed to entrench that blood-bought liberty in all its fundamentals. Thomas Jefferson wrote, "If people let the government decide what foods they eat and what medicines they take, their bodies will soon be in as sorry a state as are the souls who live under tyranny". It is sad that those who crafted the American Constitution did not, in some respects, pay better attention to Dr. Benjamin Rush—physician, educator, writer, patriot leader, and a signer of the U.S. Declaration of Independence (1745-1813). He wrote, "Unless we put medical freedom in the Constitution, the time will come when medicine will organize itself into an undercover dictatorship. To restrict the art of healing to one class [of men] and deny equal privileges to others will constitute the Bastille of medical science. All such laws are un-American and despotic and have no place in a republic. The Constitution of this republic should make special privilege for medical freedom as well as religious freedom."

Dr. Douglas Brodie is a state certified physician in both allopathic and homeopathic medicine and author of *Cancer and Common Sense, Combining*

Science and Nature to Control Cancer, he writes, "We are now reaping the bitter harvest of this omission from the U.S. Constitution. We do, indeed, now have a medical dictatorship in this country—an elite group which seeks to exclude all others in the healing arts who do not conform to the dictates of that group. Thus a form of a 'caste system' has developed, with the privileged class now using police powers to enforce its authority over those who dissent. This is the inherent nature of a totalitarian system."[8] (2005)

While the freedom of choice is certainly a freedom to fight for, when it comes to disease and its treatment we need to realize that in most cases the cause of disease is primarily a result of our bad choices. We get sick because we violate some rule of health, and like the ancient Hebrews we are "bitten" by disease. But there is a better way of treating disease than returning to the pharmacopoeia of Egypt.

God's Remedies

If one has to be rich to be healthy, then God only loves the wealthy. If the means to achieve and maintain optimum health is within the reach of only a select few, then God is elitist. But the Bible maintains that God is "no respecter of persons".[9] If so, it behooves God to make His healing available across the spectrum of society. If so, we would first expect God's remedies to be simple. King Hezekiah was told to put his house in order because he was going to die from his disease. But he wept sore and pleaded with God to spare his life. His petition was heard. He was commanded to place a poultice of figs upon his boils. He was healed and lived for another fifteen years, without the poisonous side effects of drugs.[10] **Simple.**

Secondly, God's remedies will be accessible. An early experience in the Biblical history of the Hebrew exodus from Egypt, illustrates this principle. After their triumph over the mighty armies of Pharaoh at the Red Sea, Moses leads the people out into the parched wilderness of Sinai. Ahead lay the green oasis of Mara. Moses, who had shepherded his flocks in that area for years, no doubt knew that the waters of Mara were, as the name implied, so bitter they were undrinkable. On arriving and tasting the water, the people thronged Moses, disparaging his leadership, murmuring until they convinced themselves of their certain doom. However, Moses is directed to a certain tree, "which, when he had cast into the waters, the waters were made sweet"[11] Moses was not sent on a long journey back to the sorcerers of Egypt, or on to the drugstores of Babylon, but was pointed to the remedy right within reach. **Accessible**.

Thirdly, God's remedies should be free or affordable. How often what we pay for and what we actually get are two different things. How often we spend our money only to realize we have been legally robbed. It's like enriched white bread. The food companies take whole wheat flour, mill off twenty-three vital ingredients, add seventeen back into the empty white flour, call it "enriched", and sell off the other ingredients for a tidy little sum. On the other hand, God invites all, "Every one that thirsteth, come ye to the waters, and he that hath no money; come ye, buy, and eat; yea, come, buy wine and milk without money and without price. Wherefore do ye spend money for that which is not bread? And your labor for that which satisfies not?"[12] In this passage the prophet Isaiah addresses man's propensity to ignore the truly valuable yet inexpensive necessities of life while wasting their precious earnings to gratify some transient fancy.

Jesus told the story of a Jewish man waylaid by robbers and left for dead on the road from Jericho to Jerusalem. Along came a couple other good Jews, who quickly pass by on the other side fearing danger. Next, a Palestinian man on his donkey rides up. After assessing the injuries, he pulls from his first aid bag some wine and olive oil and dresses the man's wounds. He then medivacs him to the nearest hotel where he pays for his food and lodging. Free for the Jew and **Affordable** for the Palestinian.[13]

Another vital ingredient of God's remedies is faith. Jesus plainly told those healed by His gentle touch, "Thy faith has made thee whole". Thankfully, we don't need a lot. "If ye have faith as a grain of mustard seed, ye shall say unto this mountain, Remove to yonder place; and it shall remove; and nothing shall be impossible unto you."[14] Big things grow from little beginnings. If we hold to our belief, we can move mountains of sickness, even if it takes one shovel at a time. **Faith.**

Lastly, God's remedies will be easy. There is the example of Nahmaan, the commander-in-chief of the Syrian army. He was afflicted with leprosy and was told if he went down to Israel they could cure him. When the prophet tells him to go down to the River Jordan, he is indignant, thinking he deserves some real magic for his real disease. But, when he finally humbles himself and dips seven times in the muddy waters, he comes up with the flesh of a newborn.[15] Then there is the story of Jesus Himself healing. "He anointed the eyes of the blind man with clay, and said to him, Go, wash in the pool of Siloam. He went his way therefore, and washed, and came seeing."[16] **Easy.**

SAAFE. Here we have another acronym for God's remedies. **S**imple, **A**ccessible, **A**ffordable, mixed with **F**aith, and **E**asy.

War

In 1971, by U.S. presidential decree, "war" was declared on cancer. Over the next twenty years, patients were subjected as guinea pigs to hundreds of so-called cancer-cure drugs and therapies. Yet from 1978 to 1991, the rate of cancer climbed by an alarming 18.6% for men and 12.4% for women.[17] With these kinds of "casualties" from "friendly fire", it would have been healthier to surrender. What were lauded to the heavens as great steppingstones to better health, have in fact become tombstones. As we saw in the last chapter, for millions, drugs and various therapies have been the premature cause of death.

Many caregivers today, associated with the health industry, are caught in a moral ethical war. One disenchanted medical student, upon graduating with high honors from a premier American medical school, gave his diagnosis of modern medicine as it is generally practiced: "Medicine is as distant to healing as prostitution is to love."[18] In the introduction to his book *Heart Failure—Diary of a Third Year Medical Student*, Michael Greggor, MD, quotes a passage from English poet, novelist, and critic, Thomas Blackburn:

> "This is the School of Babylon
> And at its hand we learn
> To walk into the furnaces
> And whistle as we burn."

Most healthcare providers have no misgivings about the mercenary nature of the drug cartels or healthcare systems. They are intimately aware of the poisonous nature of drugs and the great toll they are taking on human lives, never mind the economy. But they still see a great need of ministering to the sick who are daily growing in numbers. What alternatives do they have?

Some have reformed their thinking and their practices. Some, in trying to justify the exceptional use of poisonous drugs, fight a losing battle against professional, medical, insurance, and hospital protocols that make drugs more and more the rule. Others, by making blistering attacks on health practitioners of a hundred years ago, look to somehow justify their association with a modern system of sorcery that has a strangle hold on society. They vainly try to distance the drugs and practices used today from those used a century ago.

I certainly do not want to belittle the great achievements in the field of health. But, it is hard to imagine how defenders of modern medicine, in the face of current iatrogenic death statistics, fail to recognize that similar practices used in the Dark Ages are still being promoted today, only under more sophisticated labels. While a knife is no longer widely used to bleed people to thin their blood, in its place drug companies have concocted poisons to thin the blood internally. How can some berate the torturous treatments given a hundred years ago, and ignore the effects of radiation or chemotherapy on patients in our day? As for drugs, are they really more benign? They are not. Their chemistries may be different, their dosages may be more precise, but their natures are still poisonous and their long-term consequences just as cruel. The difference is like comparing large poisonous snakes with smaller poisonous snakes. One nurse, contrasting some of the drugs used a hundred years ago to those used today, commented, "The barbarism of chemotherapy would make arsenic jealous."

Arsenic Trioxide

Arsenic is a prehistoric remedy used by the medical profession for hundreds of years, until it was exposed for the evil it is. However, in recent years inorganic arsenic trioxide has been used in China in the treatment of acute promyelocytic leukemia (APL). Knowing that arsenic causes damage to the liver, bone marrow and skin, the FDA nevertheless approved arsenic trioxide (under the trade name Trisenox®) for relapsed or refractory APL in September, 2000. Today a whole new line of organic arsenical drugs is being developed for cancer therapies and an unsuspecting public.

For those who eat chicken they might be interested to know that along with antibiotics, arsenic is also included in poultry feed supposedly to protect their health and yours.[19] Laws have been passed to ban arsenic in wood preservatives. How is it the pharmaceuticals are allowed to include it in oral drugs?

Mercury

Mercury was used a hundred years ago to treat the sick. Who will deny mercury is extremely poisonous? But today it is prescribed daily under a variety of obscure names. For example Thimerosal® was first marketed in 1930, and continues to be the most widely used mercury-based preservative in vaccines. It is present in over fifty licensed vaccines. The Food and Drug Administration (FDA) recently acknowledged that in the first six months of life most children get more mercury than is considered safe by the Environmental Protection Agency (EPA). The truth is that sometimes kids go to their doctor's office and get four or five vaccines at the same time.

In his opening statement to the U.S. House Committee hearing on mercury and medicine on June 18, 2000, Congressman Daniel Burton (Indiana) stated, "My grandson received vaccines for nine different diseases in one day. He might have been exposed to 62.5 micrograms of mercury in one day through his vaccines. According to his weight, the maximum safe level of mercury that he should be exposed to in one day is 1.51 micrograms. This is forty-one times the amount at which harm can be caused." Burton's grandson, who was healthy before he received the shots, now suffers from autism. In 2004, only after years of concerted public outcry, was mercury banned from pediatric vaccines.

There are other books that deal with amalgam dental fillings and their contribution to mercury poisoning.

Antimony

Antimony is another member of the family of heavy metal poisons that was for centuries so widely endorsed. Its toxic effects are similar to other heavy metals like mercury, arsenic, lead, and bismuth. Antimony continues to be prescribed by doctors today. It is now used to treat cardiovascular disease and parasitic diseases such as Leishmaniasis, and Bilharzias.

Short-Term Memory

A hundred years ago strychnine was a favorite drug among allopathic doctors. In spite of its extremely poisonous nature it was prescribed for all manner of complaints including insomnia and, if you can believe it, morning sickness in women. Due to public pressure, strychnine fell out of common use by doctors in the 1930s. But recently it has been used in the treatment of nonketotic hyperglycinaemia (a rare metabolic disease in Hong Kong) as well as in neurological research with a new mysterious name—*strychnine neuronography*.

Do you remember thalidomide? In the 1960s doctors were prescribing it for such things as insomnia and for morning sickness. Women who took the drug in early pregnancy gave birth to children with severe birth defects, such as missing or shortened limbs. Shortly after the birth defects were observed, thalidomide was banned worldwide. However, like strychnine, the therapeutic use of thalidomide has been resurrected. It was re-approved in 1998 by the FDA to treat leprosy. The public has been reassured that sufficient safeguards are in place to protect the consumer. It would seem the pharmaceutical companies and doctors are gambling that people have short memories.

Exposed and vilified by doctors, researchers, writers, politicians, philosophers, the courts, and the media, the drug companies, to divert public attention, lobby for more controls and better standards. Consequently, the drug cartels succeed in strengthening their grip on healthcare systems as they handcuff their competitors. Meanwhile they continue to try and distance themselves from their ancestors. They claim they are not like them, that science and pharmacology are greatly advanced and today they have developed many much less toxic drugs than those employed during the previous centuries. Is this fact or fiction?

George Washington

George Bernard Shaw (Nobel prize for literature) once noted that, "A doctor's reputation is made by the number of eminent men who die under his care." Without pointing fingers let's take a look at this modern proverb in light of modern drugs.

On Friday, December 13, 1799, President George Washington was exposed to a cold rain. That night he developed severe flu-like symptoms. Washington was a firm believer in the benefits of bleeding—a common practice of the day for treating disease. Less than twenty-four hours later he was dead. He was bled of close to half his blood, given calomel (a powerful mercury-based drug producing vomiting), and blistered—the three main treatments. One who reviewed his case summed it up as "little short of murder". Now Washington's treatment was clearly the cause of death.

Like those of two hundred years ago, today's newer drugs bear in their molecules the same poisonous nature. Today's drugs may be more pure and more precisely prescribed in smaller dosages, but they can be proportionately as toxic as those drugs in Washington's day. The effects may be as immediate and profound or as delayed.

Dwight Eisenhower

On Thursday, September 22, 1955, President Dwight Eisenhower had for breakfast: sausage, bacon, mush, hotcakes; for lunch: hamburger with raw onion; and for dinner: roast lamb. He complained later that the onions gave him "indigestion". At 2 a.m. he awoke with chest pains. His personal physician was summoned and he was given morphine. He was later moved to a hospital, diagnosed as having a heart attack, and given heparin, a very potent drug to thin the blood. His long-term treatment included a low fat diet and a full month of complete bed rest (the norm was six months). To this was added a relatively new anti-coagulant drug (anti-clotting, blood-thinning) Coumadin®. It was its celebrity connection to Eisenhower that propelled Coumadin to the head of the pack.

Fourteen years later on March 28, 1969, Eisenhower died. His heart, "scarred by **seven** attacks and weakened by recent episodes of congestive heart failure, finally gave out despite the best efforts of medical science to prolong his life."[20]

Dr. Thomas W. Mattingly was Eisenhower's cardiologist and chief cardiological consultant at the time of his first attack and later death. His conclusions from reviewing Eisenhower's autopsy and records were that "Eisenhower's anticoagulant therapy [including Coumadin] was not managed as carefully as the situation required."[21] In other words, Eisenhower's life was cut short by his drug therapy. Today, the cause of his death would have been labeled as iatrogenic—mismanagement of properly prescribed drugs—which, as we saw in the last chapter, is the fourth leading cause of death in America.

Warfarin/Coumadin

Coumadin's history dates back to rotting hay on Wisconsin farms in 1921. An offending agent in the hay was blamed for the dramatic numbers of death in cattle.

The chemical was later isolated and sold as rat poison under the name of Warfarin. This next bit of information is not included for the benefit of rats.

Warfarin works with rats better than strychnine because rats are 'bait shy'. When rats see another dead rat laying so close to strychnine-laced bait they say, "Not me!" Now rats are smart, but not that smart. A rat that dies three days after eating Warfarin from internal hemorrhaging, doesn't leave a last will and testament saying, "This is the consequence of that great meal I had on Tuesday". So, as two of his buddies are munching on Warfarin-laced bait, they are talking about poor old Charley who died last week: "Stress probably." "That cat." Or, "These things just happen, and so young too." But they never connect that he dined on the same bait they are eating. It is lost on them. So, for decades Warfarin has been used as a rat poison. Rats are not salient beings that can reason from cause to effect over time.[22]

Rats are not salient beings. In other words, not the brightest light bulbs on the planet. But what does that say about humans who continue to be led like sheep to the slaughter? How Warfarin's slow-acting poison fools the short-term memory of rats, the pharmaceuticals are gambling the long-term effects of Coumadin and other poisonous drugs will fool humans.

I am reminded of something Winston Churchill said about truth: "Man will occasionally stumble over the truth, but usually manages to pick himself up, walk over or around it, and carry on." With centuries of negative history stacked up against conventional drug therapy, it is indeed marvelous to see how casually the silent majority skips over each new revelation of drug intolerance/toxicity/mortality as they head for the nearest drugstore. Churchill added, "The truth is incontrovertible. Malice may attack it, ignorance may deride it, but in the end, there it is." While the pharmaceuticals companies refine their propaganda against alternative medicine, and while some ill-informed professionals still belittle natural remedies, a week hardly goes by but there is not some new public outcry or class action lawsuit by some vocal minority "bitten" by some "new" "improved" "wonderful" and reputable drug.

I am often reminded by advocates of drug therapy that many drugs have a natural origin, that is, they are derived from plant or animal sources. But attaching the word 'natural' to some medicine does not automatically make it benign. If in its 'natural' state a plant has a reputation of being poisonous, how does extracting the poison and concentrating it make it less poisonous? It does not make it less poisonous, it makes it more poisonous. Some would even argue that, because of its connection to hay, Coumadin has a "natural" genealogy. This is a classic example of the smoke and mirrors used to justify or pasteurize the promotion of poisonous drugs. Finding it in a heap of rotting hay does not mean it is any more a natural remedy in the classic sense than strychnine, which is extracted from the seeds of the *Strychnos nux vomica* tree.

Coumadin is first and last a poison that promotes hemorrhaging internally and externally. It is lethal to cows, rats and humans. It kills outright at high levels. At low levels it seriously attacks the body's chemistry and circulatory system. What strychnine can do biochemicaly in the short and long-term, Coumadin can also imitate. In spite of all the warnings the body gives that the drug is a poison and is sick of it, the poor uninformed misguided sick patient is told by his or her concerned physician, "Expect to be on this for the rest of your life". Sounds profitable if not Machiavellian.

Are there alternatives to poisonous blood thinners? Yes! Coumadin is used to thin the blood, but there are much safer ways to do that without poisoning oneself. Water, a raw food or juice fast, a fat-free diet, to name a few. Coumadin depletes Vitamin K from the body. Vitamin K is essential for blood clotting. Patients on Coumadin are routinely told to avoid foods high in Vitamin K, which will compromise their drug therapy and put them at risk of congestive heart failure. But, we know a lot more about Vitamin K than forty years ago. Recent research has identified new benefits from its use. It may be a key anti-aging vitamin and it may prevent both heart disease and osteoporosis. It's also a stronger antioxidant than Vitamin E or Coenzyme Q10. In addition, Vitamin K may be a key feature in the future treatment of certain kinds of cancer. Lack of it may have something to do with Alzheimer's disease.

In Eisenhower's case, food and lack of physical exercise were principle causes of his heart attack. If the high fat, high meat menu Eisenhower enjoyed on the day of his first heart attack was typical of his diet, then it was a diet low in Vitamin K and other coenzymes. Because of its antioxidant benefits to the circulatory system, we can speculate that the lack of Vitamin K may well have been a contributing factor to his heart disease. The irony is that he was then put on a poisonous drug which further lowered his Vitamin K. Complete bed rest did nothing to improve recovery.

While this may all be new research, it is old knowledge. Hippocrates understood the primary importance of food 2,400 years ago when he said, "Let food be thy medicine, and medicine thy food". In our day, Thomas Edison said, "The doctor of the future will no longer treat the human frame with drugs, but rather will cure and prevent disease with nutrition."

The Hebrews were blessed with this understanding long before Edison or Hippocrates, but they never became the beacon of light God intended them to be. Unfortunately the medical profession of today has yet to embrace this new old "light."

Charcoal and Heart Disease

Incidentally, charcoal is the antidote of choice for all rat poisons. Like Moses' staff, the humble charcoal stands invincible against all the hordes of poisonous drugs. Charcoal has more than proven its ability to swallow up most any poisonous modern drug thrown its way. And, as we have already noted (Chapter 7) charcoal adsorbs excess blood cholesterol, LDL, and triglycerides, which significantly increase the risk of heart attack or stroke. In studies conducted in 1926, 1976[23], and more recently in 1986[24], cholesterol levels in blood were reduced by as much as 43% and triglycerides by 76%. This promises some immediate benefits to heart attack victims. However, low levels of blood fats returned to previous high levels within four weeks after stopping charcoal, indicating once again that charcoal does not replace a good program of total health. It would be poor judgment on the part of those who suffer from heart disease, to take charcoal to lower blood fats without simultaneously adjusting their fat intake and exercise.

In the Bible, God promised the Hebrew nation that "none of the diseases"[25] common to ancient Egypt would befall them if they obeyed His laws. These laws not only included the moral, ceremonial, and civil laws, but also the health laws. These advanced dietary and hygienic rules are considered progressive even in our times. While God has strategically placed **SAAFE** remedies within our reach to meet the varied circumstances and needs of the sick around the world, they do not replace the physiological laws of health that He has digitized on every cell of our being.

We do well to use God's **SAAFE** remedies in place of poisonous pharmaceuticals. We do better to employ the **NEW START** doctors of health to deliver us from slavery to sickness and disease. The sorcerer's enchantments were no match for the plagues of ancient Egypt, nor have modern drugs been able to check the modern plagues of chronic and degenerative disease. But, the promises of health given to the Hebrew nation long ago are still in force today for those willing to comply with the conditions. And, for all who realize their need of help, God's invitation is still open.

**"Come unto Me all ye who labor and are heavy laden,
and I will give you rest."[26]**

CR

[1] Exodus 6-14
[2] Herodotus, Histories 2, 84
[3] http://nefertiti.iwebland.com/timelines/topics/food.htm
[4] http://nefertiti.iwebland.com/timelines/topics/medicine.htm#rem30
[5] Exodus 15:26
[6] Numbers 21:4-9
[7] Liddell-Scott-Jones Lexicon of Classical Greek
[8] http://www.drbrodie.com/beliefs.shtml
[9] Acts 10:34
[10] 2 Kings 20:1-7
[11] Exodus 15:23-25
[12] Isaiah 55:1
[13] Luke 10:30-35 Author's paraphrase
[14] Matthew 9:22; 17:20
[15] 2 Kings 5:14
[16] John 9:6,7
[17] Brodie, W Douglas, MD, *Cancer and Common Sense Combining Science and Nature to Control Cancer*, WINNING Publishing, 1987
[18] Gregor, Michael, MD, *Heart Failure - Diary of a Third Year Medical Student*, Booklocker, 1999
[19] Silbergeld, Ellen K, *Environmental Health Perspectives*, May 3, 2004
[20] Felix Belair Jr., Dwight Eisenhower's Obituary, *The New York Times*, March 29, 1969
[21] Dwight D Eisenhower Library, Abilene, Kansas Mattingly, Thomas W.: Medical History of Dwight D Eisenhower, 1911-1987 http://www.eisenhower.archives.gov/listofholdingshtml/listofholdingsM/MATTINGLYTHOMASWMedicalHistory19111987.pdf
[22] Joseph M Newcomer, PhD, Computer Science, Carnegie-Mellon University, http://www.flounder.com/warfarin.htm
[23] Cooney, *Activated Charcoal in Medical Applications*, pp. 551, 456-460
[24] *The Lancet*, 2(8503):366-367, 1986
[25] Exodus 15:26
[26] Matthew 11:28

Chapter 22 Consumed by Fire

> "It is only when ignited and quenched that charcoal itself acquires its characteristic powers, and only when it seems to have perished that it becomes endowed with greater virtue."
> —Pliny A.D. 50

"Ignited ... quenched ... perished ... endowed." Charcoal, as Pliny observed long ago, is an enigma, a seeming contradiction, a riddle, a puzzle. How is it that charcoal so positively impacts our personal health, the health of our children, of our pets, and of our environment? Is it mere chance that it is so universal, so common, so available, so affordable and so safe and easy to use? As I have traveled, and as I have been called on to minister to some of the sick I meet, I have come to the conclusion that charcoal is no accident. It is so perfectly suited to every class of people, every geography, and every climate that it seems to have been handcrafted with man in mind. Its history is entwined with man's history. If inanimate nature could have a destiny, then charcoal would include man's. It is profoundly simple, but within its simplicity I see a larger picture, an object lesson, a spiritual lesson.

In the physical realm, charcoal only helps in restoring health if it is used in conjunction with good, common sense health practices such as NEWSTART®. Without the interaction of these other simple health principles, charcoal will only give a fraction of the physical benefit that it otherwise could. Charcoal also cleanses the blood of poisons and impurities that cloud or drug the mental powers, but here again, without the complimentary eight doctors of health, the benefits to the mind are only temporary. When the mind is strong, the mental powers assist in overcoming sickness. When our thinking is wrong, often our feelings follow suit. Some researchers believe that fully 80% to 90% of all disease originates in the mind. When the citadel of the mind suffers a breach in its own defenses, the trickle-down effect of negative thinking quickly undermines the physical health. Total health, vibrant health, must include mental health as well as physical health.

Philosophy of Health

Over these past twenty-seven years since I first wandered into that little health clinic in Guatemala, I have been assembling a personal philosophy of health.

Sometimes I add to it, and sometimes I have to relinquish some long cherished notion. Beginning with a flat-earth concept of health, it has slowly taken shape into a living dynamic.

My first reality was the physical—the wants and desires of the flesh. My world was flat and small. Reading and education began to tickle the dendrites of my mind, but it was nature that exposed me to the possible dimensions of my thoughts. My world took on breadth. However, it wasn't until I was all alone by a crackling fire under a clear fathomless desert night sky that it dawned upon me that there was another dimension, a depth and a height that only eternity could measure.

The discovery that I was a spiritual soul guarded by an intellect and housed in a physical body, was at first very thrilling, but also terrifying. Looking up, I found my little self sinking into illimitable space, the letters of my mind plummeting into emptiness, the atoms of my soul dissolving into mesons and quarks, and I quickly back paddled to my little fire. While I was struck with the potential of what I was, at the same time I was made keenly aware that I was mortal!

Here I was a three-dimensional, a micro trinity, encompassing the physical, the mental, and the spiritual. As I began to explore more and more of who and what I was, inevitably I was drawn back to my grass roots, my state of physical health. I noticed that when one dimension was energized it tended to buoy up the other two. The opposite was equally true. If I suffered, I suffered all together. The empathy between each of my three dimensions was so intimate that they responded as one. It became more apparent to me that I functioned best when I functioned as a whole, when each aspect of my being was free from its own version of pain or stress.

With time, I have realized that what I am able to "see" of my outer "persona" through my physical senses, I am able to apply to those inner dimensions I cannot see. The physical attributes of my health have their mental and spiritual parallels. It dawns afresh on my mind that the better I understand my physical health the better I will understand my mental and spiritual health. If I want to maximize my mental and spiritual health, then I need only apply the same principles that I employ in the physical.

My physical muscles need exercise or they atrophy. I see the same is true for my mental and spiritual muscles. My body takes a variety of simple foods and from that, biochemically meets all the requirements for physical health. The same principle of simple words, sentences, and thoughts applies to my mental and spiritual creativity. My body needs rest and recreation. So do my mind and spirit. My body needs indigestible roughage to help with digesting essential nutrients. So do the bowels of my mind and spirit need mental and spiritual "roughage". To have good strong bones they require weight-bearing stress. So do the supporting pillars of my mind and of my spirit. My body needs water to cleanse, to refresh, to renew, to re-energize. What is it that waters the mind and soul, I wondered? Whatever it is, it is clear that both my mind and spirit need regular refreshment and cleansing, just as surely as does my body.

The human body produces wastes that need to be washed away. Sometimes foreign substances invade the body poisoning it, and again the body needs cleansing. But, there are also other less tangible things like pain that also need to be "washed" from the body. How does the body deal with pain? Are the mind and spirit "washed" of their pollutants in a similar way, I wondered?

Pain, Stress, and Guilt

Physical pain has its counterpart in the mind and in the spirit. Pain is to the body as stress is to the mind, and guilt is to the spirit. Just as physical pain, or any of the symptoms we generally associate with sickness, is to our body, so is distress, or any of its dissonant cousins, to the mind. Physical and mental pain in turn mirror what guilt or its mutations are to our spiritual nature. When pain slices across the body, we react by trying to remove or relieve that pain, and rebuild a healthy body. When stress numbs the mind, we try to remove the stressors or relieve the tension, and revive the mental powers. Just so, when the spirit grieves with guilt, we long to have that guilt taken away and a sense of peace restored.

In pursuing total health, vibrant health, it is becoming clearer to me that the body may be free of disease, and the mind may be free of distress, but this happy coexistence will not last for long if the spirit is not free of guilt. Pain, depression, and guilt are symptoms. It is the symptom that tells us there is an underlying sickness that keeps the whole person from experiencing maximum health.

Trying to drug the body to cover the sensation of pain does nothing to relieve the cause. Trying to drug the mind with chemicals to suppress negative thoughts does nothing for the fire that smolders on. Just so, many unwittingly try to "drug" the spirit to mask the feelings of guilt, but the spiritual pain continues to gnaw away at the soul. Physical health is incomplete until the mental health is attended to. Too often spiritual health is treated as if it is totally independent of the body and mind.

I have chosen to close this book with a look at our spiritual well being. Charcoal, as an object lesson, provides us a tool for better understanding how we can individually experience and enjoy complete spiritual health as it relates to total health.

Charcoal and Forgiveness

Water, food, exercise, and rest each play a vital part in achieving and maintaining optimum health, but sometimes the unforeseen overtakes us, and we suffer. The body is borne down with some sickness, and it has insufficient resources to recover. What is needed is an outside influence to remove the poisons, and to tip the balances in favor of recovery. Perhaps we are suffering with a badly infected wound. So, we apply a simple charcoal poultice to adsorb any toxins, disinfect the wound and promote healing. This same model can be applied to our spiritual health.

At first, we pass along innocently through life, oblivious to the notion of guilt or remorse, until one morning we wake up with a load upon our spirit that we have never felt before. Knowing something has changed, that our spirit is not free as it once was, but not knowing how to shake this defiling poison from our heart, we struggle to our feet and bend under our new load. As long as our physical and mental health stay strong, we do okay. But if we add some new load of guilt, if our physical health stumbles, if our mind is overworked, suddenly the burden of spiritual sickness overwhelms the whole person. Crippled, we fall under our accumulated burden of spiritual poisons. What we need is some outside influence to remove the load of guilt upon the heart, disinfect the spiritual wound, and promote regeneration. If we have offended another, what we need is their forgiveness, but what if they will not give it? What if we have…?

As we have seen again and again, charcoal adsorbs to itself poisons that compromise our physical health as well as our mental health. Wouldn't it be wonderful if we had a spiritual charcoal that could instantly adsorb all our guilt, all our regret and remorse, all our shame, our fear of retribution, and leave us with perfect peace, the best ingredient in total health? Well, there is such a remedy.

There is a well-known story, set in Spain, that tells of a teenager who became estranged from his father and ran away. The father, after years of anguish, goes in search of his son Paco, (a very common name for boys of the time) and ends up in the capital city of Madrid. After a fruitless search, the father falls on the idea of running an ad in a Madrid newspaper. It reads simply, "PACO MEET ME AT HOTEL MONTANA NOON TUESDAY ALL IS FORGIVEN PAPA". The next day when the father arrives at the square, there are eight hundred Pacos waiting to be forgiven by their fathers.[1]

Guilt, shame, remorse, estrangement, how many multiple thousands of "Pacos" today live with a spiritual heart sickness, longing for what no earthly medicine may offer—forgiveness. Not just an earthly pardon, but also a heavenly forgiveness.

Burnt Offerings

In the ancient Hebrew ceremonial practices, a flawless lamb, in perfect health, was presented as a burnt offering for the sins of some penitent. The guilt-stricken person confessed his sins over its head, transferring them to the lamb. The sinner was symbolically cleansed from his crimes as the lamb was consumed by fire.

Though the lamb was innocent of the poisons of sin, nevertheless, by faith, the penitent believed the lamb took his sins to itself. The charred remains spoke to the penitent of sins consumed and cleansed by fire.

It was believed that the innocent sacrifice foreshadowed the "Lamb of God" which would come to bear the sins of the world. The penitent believed by faith that, in some mysterious transaction, God was able to cleanse them from the defiling poisons of guilt, shame, remorse, and fear of retribution, that gripped their hearts. If the guilty sinner left the sacrificial ceremony free from the burden of his guilt, it was simply because he believed by faith that he was free, and his act of faith in the promise of forgiveness had been ratified in heaven. If he left still carrying his burden, it was because he had not believed by faith. How wonderful, if only it was true.

Hundreds of years passed, and with the years, countless thousands of animals were sacrificed.

One day, two thousand years ago, in ancient Israel, beside the River Jordan, John the Baptist looked up to see his cousin Jesus walking towards him. Turning to his audience, he announced, "Behold the Lamb of God which taketh away the sin of the world."[2] Christ, the long awaited Messiah, had arrived. But, like the innocent sacrificial lamb, His life was cut short. Jesus was nailed to a wooden cross.

The Bible describes death as the wages, or the just consequence of sin.[3] Jesus, taking upon His innocence man's sins, carried the poison of man's guilt and shame and laid them in His grave. In the spiritual sense, Jesus became a sin offering made by fire. Just as adsorbing poisons to itself does not fundamentally change charcoal, just so accepting man's sins did not pollute Jesus' purity. Different Bible writers describe it this way:

> "He was manifested to take away our sins; and in Him is no sin."[4]

> "Who in His own self bare our sins in His own body on the tree, that we, being dead to sins, should live unto righteousness: by whose stripes ye were healed."[5]

> "God made Christ to be sin for us, who knew no sin; that we might be made the righteousness of God in Him."[6]

Jesus was God's universal antidote for guilt and shame and disobedience. Just as all who were bitten by poisonous snakes were forgiven and miraculously healed when they looked at the brazen serpent that Moses lifted up in the Sinai desert, so all who will look to the uplifted form of Jesus on the wooden cross will be forgiven of all transgression, and one day healed with immortal bodies.[7] Jesus is the supernatural King of remedies.

Jesus, who knew no sin, took upon Himself the sins of the world and became the blackness of sin for us, that we might be transformed by and to the purity of

God. He took on the blackness of eternal death that we might approach unto His eternal light. He experienced the flames of the second death, the unquenchable burning lake of fire that we might bathe in the river of life. Under the horrendous weight of the sins of the world, His soul opened up and drew to His spotless heart the poisons of all humanity, cleansing every believer from all transgression. He accepted our poison that we might accept His spotless purity. No magic there, just God's glorious eternal mystery!

Miraculous

When charcoal removes the load of poisons, the physical body quickly rebounds with health. When a son is forgiven and reconciled to his earthly father, a great burden is rolled away from his heart and mind. Just so, when the soul, weighed down with a guilt and grief no man-made medicine may relieve, exposes itself to the cleansing forgiveness of God, there is complete restoration.

In Chapter 8, Ruth told the story of a miraculous recovery from poisoning with pesticide. What should have been just another miserable account of suicide, or at best a life of severe physical/mental damage, was transformed into a tribute to the power of charcoal and prayer. The Great Physician blessed the science of chemistry and the prayer of faith. But what about the state of mind that drove that man to attempt suicide?

Neither the man nor his wife was a Christian when their lives fell on this crisis. But, because of the miraculous recovery, both he and his wife were moved to learn more about the name Jesus—the name invoked for healing. Touched by the good news that not only can God revive the diseased body but also the guilt-plagued soul, they chose to take the name of Christ and be baptized, signifying the death of the old, the washing away of their past sins, and the regeneration of the new life.

Though Ruth's story took place in India, it is played out a thousand times each and every day around the world as people seek to extinguish the fire that torments their hearts and minds. But, how often, when the rescuers arrive in their ambulances, drugs and technology quickly take over. Seldom if ever does the patient hear an audible prayer invoking God's healing from those who are in charge. Any reference to divine intervention is largely ignored or bypassed. Some despairing souls are rescued. Some who have poisoned themselves are saved by the tremendous detoxifying power of charcoal, but how often are they offered a remedy for their sin-poisoned hearts?

As there are **SAAFE** remedies for the body and mind, so there are **SAAFE** remedies for the spirit.

Faith and Health

There is a philosophy of science today that masquerades as health and healing. But, it is the science of greed, of power, of enchantment without substance, and everything else but the science of healing. True healing includes faith. To everyone a measure of faith has been given, and faith requires exercise just as surely as do our legs. Faith has two legs, knowledge and self-control. Faith needs both. A one-legged faith only goes in circles. A lung surgeon may know a great deal about lungs and cancer, but knowledge alone will not rescue him from lung cancer if he smokes. A young woman may have wonderful discipline over her appetite, but if she ignores all the information about her body's need for weight bearing exercise in her twenties and thirties, all her self-control will not keep her from suffering with osteoporosis in her sixties.

Are you borne down with disease? Then, whatever the affliction, turn heavenward with your call for help. Ask for the knowledge to discern the nature of the affliction. Then employ God's **SAAFE** remedies as you co-operate to bring the body back into harmony with His laws of health. With some basic practical knowledge of chemistry, physiology, and nutrition, and the prayer of faith, you may expect God's healing touch. As you exercise faith you will receive a blessing of greater faith.

When looking for remedies for spiritual disease make sure they fit the **SAAFE** rule. Jesus has actually simplified it for us. He promises, "**Ask**, and it shall be given you; **Seek**, and you will find; **Knock**, and it shall be opened unto you: for everyone that asks receives; and he that seeks shall find; and to him that knocks it shall be opened."[8] Do you need a spiritual remedy? Just **ASK**.

Jesus, though a carpenter by trade, was minister to all, high and low, rich and poor, the educated and the unlearned. He ministered to the helpless, the alone, and to those who ministered. Gentle, winsome, noble yet so unassuming that humanity barely took notice of Him, except for His miracles. His miracles ministered to the most fundamental needs, eradicated the inroads of sin, and restored a living faith in their heavenly Father. He did this by employing the simplest of methods. The minds were lifted up above the feeble efforts of man to the powerhouse of heaven, and great miracles followed. The methods employed were so simple that it was unmistakably clear that healing came from God and not from man.

Teachers

In the life of Jesus we find a working model for the modern day physician, nurse, minister, and teacher. Healing, ministering, teaching, traveling, were daily experiences for the disciples of Jesus. But it was Jesus' attention to the physical needs of His audience that were riveted in the memories of His disciples. If Jesus preached to an audience, it was only because He had first healed. And, in healing, the minds and hearts of the sufferers were drawn to the message of a compassionate and loving heavenly Father.

The commission Jesus gave to His followers is the same today as it was two thousand years ago, and it is to be attended with physical, mental and spiritual healing. While Jesus did not give countenance to drug medication, He did sanction the use of simple and natural remedies. Unfortunately, as in the general population, so in the gospel ministry, there is a strange absence of personal knowledge of God's inexpensive simple healing medicines.

Training

The word "gospel" is derived from the Greek, which means, "to announce good news". Relieving physical and mental suffering demonstrates the practical working of the "gospel". Any "gospel" worker who understands how to treat disease with nature's simple remedies will be twice as successful in his ministry. Ellen White repeated this fact over and over to the "gospel workers" within her church. "He should have a training that will enable him to administer the simpler remedies for the relief of suffering… They should be as well prepared by education and practice to combat disease of the body as they are to heal the sin-sick soul by pointing to the Great Physician."[9] In the same way, those who are not ordained ministers or licensed healthcare workers may also become medical missionaries wherever they are, whether within their own culture or far away. Christ did not feel He was infringing on physicians when He healed the sick, nor should any in His church.

As with science, there is also a spurious kind of faith. Faith often lies so close to presumption, it is sometimes hard to distinguish between them. With "In faith alone" on their lips, some will invoke the healing of God while they leave undone His "preventives". They ask the God of nature to heal while nature's laws continue to be violated. True faith will lay hold of knowledge as it employs the simple agencies of nature for the recovery of health. Ellen White warned the workers within her church, "If we neglect to do that which is within the reach of nearly every family, and ask the Lord to relieve pain when we are too indolent to make use of these remedies within our power, it is simply presumption."[10] The world today is languishing for that true faith.

Purity

The physics and chemistry of the lowly charcoal bear an uncanny similarity to the work and ministry of Jesus. Charcoal makes raw sugar cleaner and sweeter. It removes impurities from vegetable oils and makes them smooth like silk and tastier. Charcoal makes water purer and more refreshing. Charcoal takes polluted air and returns it fresher and invigorating. Mixed with other ingredients, charcoal can be an explosive power for good. All this and more, God's gift of forgiveness can do for you and me.

By His Spirit, Jesus goes to work to cleanse body, mind and spirit from the defilements of sin and make us conformable to His righteousness. He removes the stench of selfishness and the gall of bitterness. He adsorbs the poisons of guilt and eternal death that we might approach His eternal light.

How often we give praise where little or no credit is due. How often those who are healed quickly forget their sufferings and promises, and return to their indiscretions. How quickly we forget our pleadings and vows to God, and again break His laws. How often we just presume He will give healing again and again… How often we become proud or arrogant because our prayers have been answered. How often we judge others who continue to languish in sickness. How often we resist God's still small voice pleading with us to be gathered under His wings of protection, but instead we stiffly turn away.

Some poisons, charcoal does not adsorb at all, and others only poorly. Nor does God's forgiveness redeem those sins that are not confessed and forsaken. Why, you ask? It is because of the hardness of the heart. If the mind and heart refuse to acknowledge their slavery, then there is no need of deliverance. But God promises, "Humble thyself in the sight of the Lord and He will lift you up."[11]

But, before the crown the cross. Jesus was lifted up upon a wooden cross that He might become a beacon of health and healing to the world. Those who choose to follow His example of selfless service to humanity and obedience to the laws of God will also have a cross to bear. But, they also have the promise that God will transform their cross into a glorious crown.

Graphite, charcoal and diamonds are composed from the same basic carbon building blocks but they are structurally different. In its simplest form, carbon congregates in planes and sheds easily, as in soft graphite. But that same carbon, exposed to intense heat, changes into crusty but intricate, geodesic-like fragments. Again, when it is exposed to extreme heat and extreme pressure, the uninspiring foreboding black carbon is, by the miracle of God, reborn into a dazzling crystal clear diamond. Like the humble carbon, we too have a royal pedigree.

John the prophet, the last of Jesus' twelve disciples, penned these words of encouragement and inspiration to all generations: "Beloved, now are we the children of God, and it does not yet appear what we shall be: but we know that, when He shall appear, we shall be like Him; for we shall see Him as He is."[12] The prophet Paul added, "In a moment, in the twinkling of an eye, at the last trumpet… we shall be changed. For this corruptible must put on incorruption, and this mortal must put on immortality."[13]

On His cross, Christ was offered a stupefying potion to deaden the sense of physical pain. When He tasted it He refused it. He was not willing that anything should becloud His senses and jeopardize His communication with heaven. In this He has left an example and a promise that, "When you pass through the waters, I will be with you; and through the rivers, they shall not overflow thee; when you walk through the fire, you shall not be burned, neither will the flame kindle upon you."[14]

From the cradle to the grave, from the little "owie" to the crushing load of guilt and sin, God has provided an antidote for every poison, a healing balm for every wound. The mystery of charcoal's benign power to heal is but a faint shadow of God's powerhouse of redeeming love. As we see how simple charcoal and other simple remedies really are, so simple a child may use them, we are struck with the thought that we are indeed only children. Ignited … quenched … perished … and endowed, children of the King of heaven. CR

---

[1] Hemingway, Ernest, "The Capitol of The World," in The Stories of Ernest Hemingway, New York, Scribner, p.38, 1953

[2] John 1:29

[3] Romans 6:23

[4] 1 John 3:5

[5] 1 Peter 2:24

[6] 2 Corinthians 5:21

[7] 1 Corinthians 15:52-54

[8] Matthew 7:7, 8

[9] White, Ellen G, *Medical Ministry*, Review & Herald Publishing, p. 253

[10] *Ibid.*, 230

[11] James 4:10

[12] 1 John 3:2

[13] 1 Corinthians 15:52, 53

[14] Isaiah 43:2

Emergency First Aid

911 or Poison Centers

United States—Poison Control Center 1-800-222-1222 —Calls are routed to the local poison control center.

Canada—the number can be found within the front cover of your phone directory.

Animal Poison Control Center—U.S. 1-888-426-4435

Poisoning

✓ Say a prayer for wisdom, speed and calmness.

✓ Manually induce vomiting by massaging the back of the throat with a finger.

- **Do not induce vomiting** for ingestion of gasoline, kerosene, lighter fluid, or an acid or caustic agent.
- For acids, neutralize with baking soda in water.
- For caustic agents, neutralize with vinegar in water.

✓ **Administer charcoal.**

- Stir charcoal into a minimal amount of water. Using a straw, suck the charcoal far back on the tongue. Swallow quickly. Follow charcoal mixture with 2 glasses of water.
- Give as soon as possible while patient is conscious and able to swallow.
- Small children will resist and must be held. Laying a child on his back will prompt him to swallow reflexively. Use a spoon or small bulb syringe to give the charcoal mixture.
- Caution: do not administer anything if child is sleepy, fainting or unable to swallow. Seek professional emergency help.

Estimated amount of poison or medicine taken	Amount of charcoal given if no food eaten in last 2 hours	Amount of charcoal given if food eaten in last 2 hours
1 teaspoon 1-2 tablets 1-2 capsules	1-2 Tablespoons in water followed with 2 glasses of water	4-10 Tablespoons in water followed with 2 glasses of water
1 Tablespoon 3-5 tablets 2-5 capsules	3-4 Tablespoons	6-15 Tablespoons
Unknown	1-5 Tablespoons	5-15 Tablespoons

✓ **Repeat all dosages in 10 minutes, and again if symptoms begin to worsen.**

Poisonous Bites or Stings

For immediate relief:

✓ Dissolve enough powder or crushed tablets in a minimal amount of water to make enough paste to cover the affected area.

✓ Spread paste directly onto the skin.

or

For longer relief:

✓ Dissolve enough powder or crushed tablets in a minimal amount of water to make enough paste to cover the affected area.

✓ Spread on half a sheet of paper towel or cotton cloth and cover with the other half.

✓ Mold over the affected area, cover with plastic and make secure.

✓ For multiple stings, add 2 cups of charcoal powder to warm bathtub water and immerse body for up to one hour.

Jelly Poultice

✓ Grind 3 tablespoons of flaxseed (or use cornstarch).

✓ Mix flax meal together with 1-3 tablespoons of charcoal powder.

✓ Add 1 cup water.

✓ Set aside for 10-20 minutes to thicken, or mixture may be heated and allowed to cool.

✓ Spread the jelly evenly over an appropriate size cloth or paper towel.

✓ Cover the jelly with a second cloth or paper towel.

✓ Position poultice over the area to be treated (i.e., liver, stomach, kidneys, spleen, knee, eye, ear, sting or bite area).

✓ Cover poultice with plastic 1 inch larger all around (to keep paste from spreading and drying too quickly). Secure with surgical tap or ace bandage.

✓ Leave poultice in place overnight or from 2-4 hours, if applied during the day.

Plain Poultice

This poultice, without any thickening agent, is a variation of the one described above. Consequently the charcoal may dry out more quickly and will need to be changed or remoistened.

✓ Mix charcoal (1 to 2 Tbs.) with a little water to form a wet paste. It should be moist but not crumbly or drippy.

✓ Spread the paste on one half of a folded paper towel, loosely woven cloth, or piece of gauze cut to fit the area to be treated. When ready the cloth should be moist, and thoroughly saturated with the paste.

✓ Then cover the paste by folding over the other half of the paper towel or cloth.

✓ Next place the charcoal poultice on the affected body part making sure it completely covers the area.

✓ Cover the poultice with plastic (when available, plastic food wrap works fine) cut to overlap the poultice by an inch on every side. This will keep it from drying out. If the charcoal dries out, it will not be able to adsorb.

✓ Finish off by bandaging or taping the poultice securely in place. Leave it on for several hours, or better yet, overnight. After 6 to 10 hours another poultice can be applied.

Note: Poultices of any kind only work if there is continuous moist contact with the skin.

Slurry Water

To be taken internally by babies or those with sensitive digestion as in ulcerative colitis, Crohn's disease, irritable bowel disease.

✓ Stir 2 to 3 tablespoons of charcoal powder into a quart of warm water.

✓ Allow the charcoal to settle out then pour off the gray water into a baby bottle or separate glass.

✓ This can be repeated several times using the original charcoal.

Snake Bites

How can charcoal help in snakebites? It adsorbs the chemicals in snake venoms that destroy red blood cells. Swelling begins within ten minutes if the snake is venomous.

✓ Immediately wash the area thoroughly with soap and water.

✓ Submerge the area in cool water to slow the circulation of venom.

✓ Add charcoal to the water – 1/2 cup to 2 – 5 gallons of water.

✓ Leave area submerged for about 30 minutes to 1 hour.

✓ Prepare a charcoal poultice.

✓ Cover the area with a large charcoal poultice.

✓ Change every 10 to 15 minute until the swelling and pain are gone.

✓ Give charcoal orally as well.

- Take 2 tablespoons* in 1/2 glass of water every 2 hours for 3 doses.
- Take 1 teaspoonful* every 4 hours for the next 24 hours.
- Follow each charcoal dose with 2 glasses of water.

*There is no danger of taking more charcoal.

Appendix

Some Substances Adsorbed by Activated Charcoal

Acetaminophen
Aconitine
Amitriptyline Hydrochloride
Amphetamine
Antimony
Antipyrine
Arsenic
Aspirin
Atropine
Barbital Barbiturates
Ben-Gay
Benzodiazepines
Cantharides
Camphor
Carbon dioxide
Chlordane
Chlorine
Chloroquine
Chlorpromazine
Cocaine
Colchicine
Congesprin
Contac
Cyanides
Dalmane
Darvon
DDT
Digitalis—Foxglove
Digoxin
Dilantin
Diphenoxylates
Doriden
Doxepin
Elaterin
Elavil
Equanil
Ergotamine
Ethchlorvynol
Gasoline
Glutethimide
Golden Chain
Hemlock
Hexachlorophene
Imipramine
Iodine
Ipecac
Isoniazid
Kerosene
Lead Acetate
Malathion
Mefenamic Acid
Meprobame
Mercuric chloride
Mercury
Methylene Blue
Methyl Salicylate
Miltown
Morphine

Mucomyst
Muscarin
Narcotics
Neguvon
Nicotine
Nortriptyline
Nytol
Opium
Oxazepam
Parathion
Penicillin
Pentazocine
Pentobarbital
Pesticides
Phenobarbital
Phenol
Phenothiazine
Phenylpropanolamine
Phosphorus
Placidyl
Potassium cyanide
Potassium permanganate
Primaquine
Propantheline
Propoxyphene
Quinacrine
Quinidine
Quinine
Radioactive Substances
Salicylamide
Salicylates
Secobarbital
Selenium
Serax
Silver
Sinequan
Sodium Salicylate
Sominex
Stramonium
Strychnine
Sulfonamides
Synthetic Multivitamins with Mineral
Talwin
Tin
Titanium
Tofranil
Tree Tobacco
Yew
Valium
Veratrine

Also included: some silver and antimony salts, many herbicides. In total some 4000+ chemicals, drugs, toxins, and wastes. For cyanide, mineral acids, caustic alkalis, alcohol, or boric acid, other antidotes are more effective. Remember, in any poisoning emergency time is of the essence. Send a quick prayer for wisdom, administer charcoal, and, if available, consult a Poison Control Center or a doctor for instructions and information.

Further Reading

Activated Charcoal
David O. Cooney, TEACH Services, 1999

Activated Charcoal in Medical Applications
David O. Cooney, Marcel Dekker, 1995

Rx: Charcoal
Agatha Thrash, MD, & Calvin Thrash, MD, New Lifestyle Books, 1988

Home Remedies
Agatha Thrash, MD, & Calvin Thrash, MD, Thrash Publications, 2001

The Ministry of Healing
Ellen G. White, Review & Herald, 2001

Counsels on Diet and Foods
Ellen G. White, Review & Herald, 2001

Hydrotherapy video:
<http://www.ucheepines.org/country_life_books.htm>

Hydrotherapy: Simple Treatments for Common Ailments
Charles Thomas, PhD, Clarence Dail, MD, TEACH Services, 1991. Well illustrated.

Rational Hydrotherapy
John Harvey Kellogg, MD, TEACH Services, 2001
Reproduction of Kellogg's classic 1901 textbook on hydrotherapy.

Measurement Conversions

1 grain = .065 gram

1 dram = 1/16 ounce

1 gram = 0.03527 oz.

1 oz. = 28.35 grams

1 kilogram = 2.2046 lbs.

1 lb. = .4536 kilograms

16 ounces = 1 pound

1 cup = 8 ounces

3 teaspoonfuls = 1 tablespoonful

1 tablespoonful = 1 fl. ounce

1 tablespoonful = 1/2 ounce = 10 grams = 14 capsules

16 tablespoonfuls = 1 cupful

60 drops = 1 teaspoonful

2 cups = 1 pint = 0.473 liters

2 pints = 1 quart = 0.946 liters

4 quarts = 1 gallon = 3.785 liters

British dry quart = 1.032 U.S. dry quarts

10 milliliters = 1 centiliter = .338 fluid ounce

1 liter = 1.06 quarts liquid, 0.9 qt. dry

10 liters = 1 deciliter = 2.64 gallons

Glossary

Abscess – a pus-filled cavity resulting from inflammation and usually caused by bacterial infection

Amalgam – a paste of mercury, silver and tin for filling tooth cavities

Anaphylactic – a sudden severe and potentially fatal allergic reaction to a particular substance

Anthrax – a highly infectious fatal bacterial disease

Antibiotic – a substance that is able to kill or inactivate bacteria in the body

Antifungal – preventing, reducing or killing the growth of fungi

Antihistamine – a drug used to prevent allergic effects

Antiseptic – an agent that will inhibit microorganisms

Camphor – a strong-smelling compound used in medicinal creams for its antiseptic and ant-itching properties

Cellulitis – an infection and inflammation of the tissues beneath the skin

Cirrhosis – a chronic progressive disease of the liver

Compress – a cloth pad, often moistened or medicated and pressed firmly against a part of the body as a treatment

Croup – an inflammation of the larynx and trachea with cough, hoarseness, and difficult breathing

Diabetes – a condition of metabolism in which sugars and starches cannot be properly handled because of insufficient insulin

Dialysis – the process of filtering the accumulated waste products of metabolism from the blood of a patient whose kidneys are not functioning properly

Diarrhea – frequent and excessive discharging of the bowels producing abnormally thin watery feces

Dispensatory – a commentary of the medicinal substances listed in the national pharmacopoeias

Diuretic – a substance that increases the flow of urine

Douche – a cleaning of part of the body with a jet of water.

Drench – to give an animal a large dose of medicine in liquid form by mouth.

Drug – (*drugan*, to dry) material agents of every kind employed in the treatment of disease—originally applied to vegetable medicines

Eclectic – choosing what is best or preferred from a variety of sources or styles

Encephalopathy – any disease of the brain

Endometritis – an inflammation of the lining of the womb

Erythroblastosis fetalis – a condition in infants of incompatibility of blood factors, specifically Rh factors

Flatulence – excessive gas in the stomach and intestine that causes discomfort

Fomentations – the application of heat with hot packs.

Gallstones – a small hard mass that forms in the gallbladder as a result of infection or blockage

Gangrene – local death and decay of soft tissue of the body as a result of lack of blood to the area

Halitosis – bad breath

Hemoperfusion – the passage of blood through columns of adsorptive material, such as activated charcoal, to remove toxic substances from the blood

Hepatitis – inflammation of the liver, causing fever, jaundice, abdominal pain, and weakness

Inorganic – chemical compounds that contain no carbon

Intubation – the procedure of inserting a tube into the trachea of a patient who is not breathing

Morbidity – the presence of illness or disease

Mucosal – involving the tissues that line body cavities, tracts, and passages

Narcotic – an agent that promotes insensibility

Necrosis – death of cells in a tissue or organ caused by disease or injury

Nephron – the functional unit of the kidney responsible for the actual purification and filtration of the blood

Nosocomial – originating or occurring in a hospital

Organic – compounds characterized by chains or rings of carbon atoms linked to hydrogen, oxygen, nitrogen, and other elements

Palliative – alleviating pain and symptoms without eliminating the cause

Pharmacopoeia – the official list of drugs and their preparations recognized by the medical profession in a certain country

Phlebitis – inflammation of the wall of a vein

Pleurisy – inflammation of the membrane surrounding the lungs

Poultice – a medicinal mass spread on cloth and applied over sores or area of treatment

Scours – diarrhea usually only associated with incorrect milk feeding

Sepsis – the condition or syndrome caused by the presence of microorganisms or their toxins in the tissue or the bloodstream

Sitz bath – a shallow, tepid bath often recommended to soothe the discomfort and pain of conditions associated with the trunk area

Slurry – a liquid mixture of water and charcoal

Tryglicerides – a fatty particle most prevalent in animal fats

Tetanus – an acute infectious disease causing severe muscular spasms

Typhoid – a sometimes fatal bacterial infection of the digestive system

CR

Index

C

D

E

F

O

P

CR

An Invitation

If you have found a blessing in these pages, or if you have proved charcoal to be a successful remedy for yourself or others, for your pets or livestock, for your garden or field, I invite you to submit your story to our website so others may benefit from your experience.

www.charcoalremedies.com

Gatekeeper Books

USA